World of Geology

Published by
Whittles Publishing Ltd.,
Dunbeath,
Caithness, KW6 6EG,
Scotland, UK

www.whittlespublishing.com

ISBN 978-184995-437-2

Printed and bound in Great Britain
by Severn, Gloucester

World of Geology
Travels to Rocky Places

Tony Waltham

Contents

Preface

Travelling the world, always armed with a camera, led to the author compiling a substantial collection of photographs, many of them with a geological theme. They became a core for lecture presentations and university teaching, because geology is such a visual subject.

Many of the photographs then appeared alongside short texts on the back covers of the popular magazine *Geology Today*. Though originally intended to showcase material from various readers and geologists, the back cover became an item that was provided by the author for every issue. Its welcome reception by the readers led to the idea of compiling a book.

This in turn evolved into a set that includes more than those from the magazine back covers. It has grown into a worldwide overview of just a fraction of the magnificent sights, both natural and influenced by mankind, that make the geological world so totally fascinating and frequently so beautiful. These photographs cannot cover the whole world, because they are just from the travels of one person, and also because it would take a thousand photographs to cover the fabulous variety of spectacular geological features. Instead, they are offered as a taste of the visual delights within the world of geology.

The photographs and their texts just follow a grand tour across the surface of our planet, and their themes are outlined in the list of contents. Perhaps the whole book is best viewed as a glorious journey of discovery.

The many travels and the entire book have only been made possible by my wife Jan, who has long been the best travelling companion in the world and appears as scale on so many of the photographs.

Tony Waltham, Nottingham,
2019

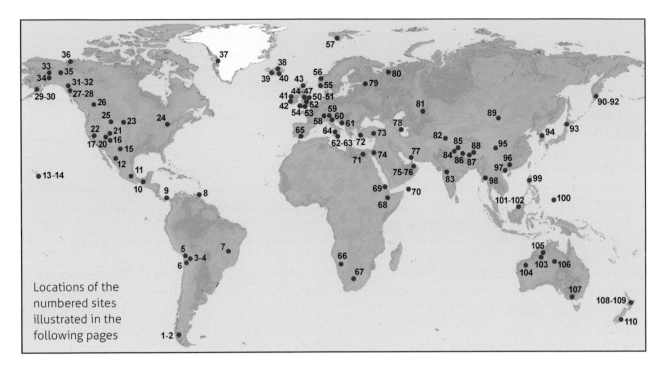

Locations of the numbered sites illustrated in the following pages

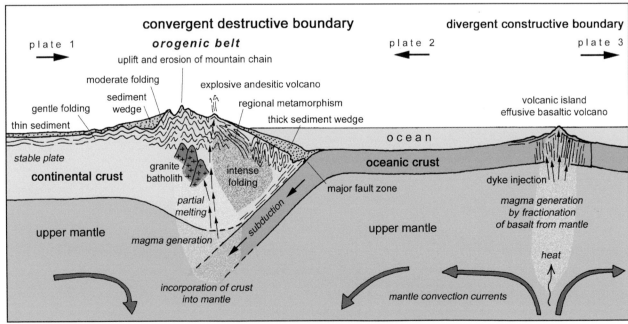

All the major geological processes related to convergent and divergent plate boundaries

Introduction

The world of geology is the world as we know it, that we see and that we live on. It is all about the evolution of the Earth's crust, the nearly rigid layer less than 100 km thick that is the outer shell of our evolving planet. This crust is broken into a few dozen large and small tectonic plates, which move around at rates of a few centimetres per year. Originally known as continental drift when it was first recognized in 1912, this geological activity has been referred to as plate tectonics since its processes began to be properly understood during the 1960s.

A large part of the Earth's crust is the oceanic floor. The basaltic rock of the slowly moving oceanic plates is continuously being created along plate boundaries that are divergent and destroyed along those that are convergent. These are the major processes of plate tectonics that keep Planet Earth evolving and alive. The oceanic basalts are similar to those in some types of volcano, but otherwise they remain largely unseen beneath the cold, dark and minimally explored waters of the ocean depths.

The second part of the Earth's crust is the incomplete upper layer, largely of granitic composition, that forms the continents. Along with the submerged edges known as the continental shelves, these occupy about one third of our planet's surface. Continental crust cannot be destroyed by subduction along the convergent plate boundaries because it floats on the heavier basalt. Instead, it is continually evolving and reshaping itself by being crumpled, squeezed, distorted, heated, locally melted, eroded and re-deposited. Most of this activity occurs along convergent plate boundaries, and accounts for the complexity and variety of the rocks that are seen across and beneath the Earth's surface.

Squeezed and crumpled continental rocks form the great mountain chains, the Himalayas, Alps, Andes and more, all developed along convergent plate boundaries. The bulk of these chains consists largely of metamorphic rocks, changed from whatever their origins by heat and pressure in the zones where plates collide. Some of their material is melted at depths and becomes igneous rock, cooling down as huge granite masses or erupting hot from volcanoes; they are matched by the volcanoes and igneous rocks that create the oceanic plates along the divergent boundaries. Rocks of the third super-family are known as sedimentary, and these formed at the Earth's surface, mainly from swathes of sediment that is composed of debris eroded from any available rock outcrops. Igneous, sedimentary and metamorphic rocks continually evolve from one to another in the great Cycle of Rocks that keeps our planet active, alive and inhabitable.

Almost a sideline within that Cycle, erosion shapes the landscapes of our world. Largely the results of water action, but with significant contributions by ice, wind and gravity, the major processes of erosion are powered by the uplift of the mountain chains. Rivers and glaciers flow from the high ground, and their movement accounts for most of the world's erosion. Large-scale development of our landscapes is one more component of plate tectonics.

The background of our world is the geology. Superimposed on it are the processes of life. Animal life within the oceans produces shell and skeletal material that is a major component of limestones formed on the seabed before being uplifted to form new land. Plant life largely feeds the animal life, but some of it, in the right deltaic environments, is preserved to form coal. Hence the world of biology is interwoven to be a part of the world of geology. The many cyclic processes of geology fit together, with the result that Planet Earth can evolve continually and survive. This is fortunate, because non-cyclic processes tend to be terminal, and, if they were dominant, would soon render the planet uninhabitable.

Geology has an important fourth dimension: time. Ever since Earth became a recognizable planet about 4500 million years ago, continuous geological processes have seen it evolve into the complex world that now exists. The oldest know rock dates back 4000 million years. All the other rocks have been transformed, melted, metamorphosed, eroded or re-deposited in subsequent phases of plate tectonic activity. Most rocks now to be seen in the ground are between 1000 million and 10 million years old, though some far older rocks do survive and new volcanic rocks are being formed at all times.

Broadly speaking, the rocks are old and the landscapes are young. Erosion shapes a landscape by working away at rocks formed long before and in totally different environments. Timescales do vary. Some parts of interior Australia have changed little in 100 million years, yet the Himalayas have developed into a giant mountain range entirely within about the last ten million years; marine limestone at the top of Mount Everest is just one consequence of the uplift generated by plate convergence. Most landscapes have evolved within the last few million years. Outside the tropics, many of their details can be traced back only to the last of the world's great glaciations around 15,000 years ago. Even younger, the latest component of geology is the work of mankind. Large-scale mining, major construction projects and vast agricultural schemes are significant components of the planet's landscapes, and most of these developments have been within the last hundred years or so. Man now counts as a significant geological process.

It all adds up to the vast and ramifying story that is geology. It may be difficult to comprehend the immensity of geological time, because it reaches back so far. But time lies at the heart of geology. It accounts for the many layers of infinitely slow or terrifyingly rapid processes that have formed the rocks and landscapes that are the world as we see it.

Granite laccolith at Torres del Paine

The southern end of the Andes mountain chain is commonly known as Patagonia, though strictly this name is that of the southern province of Argentina. The dramatic and vertiginous peaks of the Torres del Paine lie within a national park on the Chilean side of the Andes, and are matched by those of the Fitzroy Mountains in another national park, nearby but on the Argentinian side. Both groups of mountains are formed in Neogene granitic intrusions that are parts of the extensive and complex Patagonian batholith. The granites are remarkably massive; horizontal and inclined fractures are rare, and primary vertical factures are widely spaced. Consequently, glacial erosion and deep dissection of the terrain have developed huge, near-vertical rock walls; some of these completely surround spectacular rock towers that are many hundreds of metres high and far taller than they are wide.

The peaks of the Cuernos del Paine lack the locally characteristic narrow profiles, but are distinguished by their black caps above the pale-grey granite walls. The black rock is Cretaceous mudstone that has largely been metamorphosed to far harder hornfels by heat transferred from the granite intrusion immediately beneath. These caps are remnants of the original roof-rock above the intruded granite; that roof was more than a kilometre thick and the eroded remains of it extend for 20 kilometres through the mountain peaks. Granitic magma rose along the major Lago Grey fault, until the reduced overburden pressure allowed it to intrude and spread almost horizontally eastwards through the mudstone, ending up in the form known as a laccolith; this has a crudely lenticular shape with undulations in both roof and floor. The Paine intrusion occurred in three phases, over a period of about 90,000 years ending about 12.5 million years ago, with each of the two later phases intruded beneath its predecessor. Wherever exposed, both the upper and lower margins of the laccolith are sharp and clean. This indicates that the intruding material consisted of mature granitic magma, and was not just a silica-rich fluid that invaded the country rock to generate *in-situ* meta-granite. The towers of Paine do expose a most striking and instructive example of a granite intrusion.

Perito Moreno Glacier

Named after a noted explorer and conservationist, the Perito Moreno is surely one of the world's great tourist glaciers. Near the southern tip of the Andes in South America, it descends from the Southern Patagonian Icefield, eastwards into Argentina, and ends in the country's largest lake, the multi-armed Lago Argentina (which is impounded by a Quaternary moraine out on the margin of the Patagonian pampas). By a quirk of good fortune, the glacier ice extends across the lake and almost reaches the opposite shore, where a rocky headland provides perfect viewpoints from a cascade of boardwalks. A visitor centre has easy road access from the nearby town of El Calafate, and the entire site certainly draws in the crowds; that town has become a regular stop on the Patagonian tourist trail largely on account of this one incredibly beautiful and readily accessible glacier.

With a daily advance of nearly two metres into the open water, the blue ice wall at the glacier front calves frequently, with chunks, slices and pinnacles of ice crashing into the lake in dramatic style. Its scale is difficult to appreciate, but the ice wall rises nearly 60 metres above the water, and the far side of the lake is more than two kilometres away. Besides floating off into the lake, the glacier is permanently creeping towards the rock slope below the walkway (off to the left of this view). When, occasionally, it does lie right against the rock, it prevents the southern arm of the lake draining past the ice and towards the lake outlet (far away towards the right). In such situations, the water level in the confined southern arm rises by as much as ten metres, until the ice dam is breached in a chaos of eroding crevasses, merging ice-tunnels and collapsing seracs. This has happened 18 times since 1917, and the breakthrough events can be spectacular, though seeing one is down to pure luck, because their timing is unpredictable. However, each event leaves behind it an ice gorge that lies just below the boardwalks, and that is always there and in full view at this remarkable site.

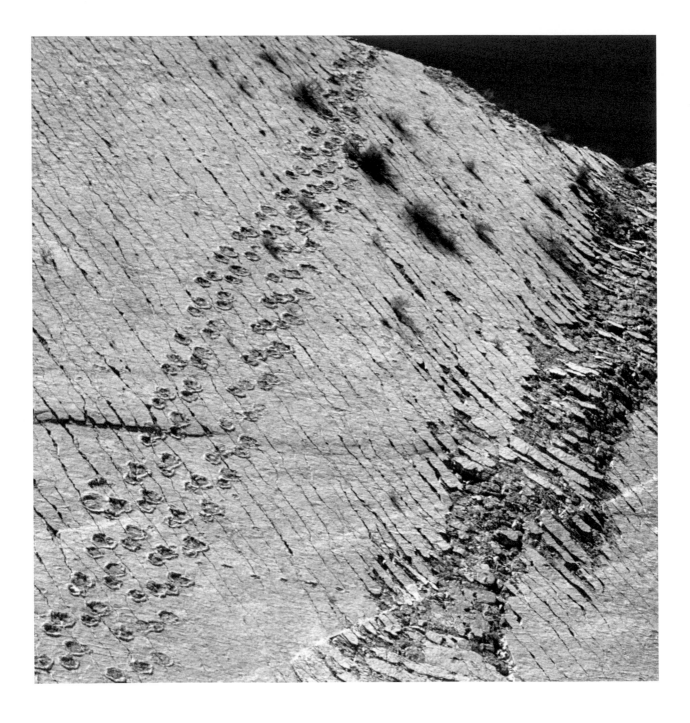

Dinosaur tracks

Just outside the city of Sucre, in the central highlands of Bolivia, a cement factory at Cal Orcko is supplied by a massive quarry that has exposed a spectacular fossil site. Limestones and clays were deposited in an Upper Cretaceous lake, and subsequent folding saw the thinly bedded sequence tilted to a dip of about 70°. Quarrying of the best limestones has left an exposed face more than 400 metres long and over 50 metres high. Most of this face is a single bedding plane on the top of a single carbonate bed about 100 mm thick, which has an equally thin clay behind it. It appears that the shallow lake was a popular spot for dinosaurs, which left more than 250 trackways indented into its soft sediment floor. These are now beautifully exposed on the wide dipping slab. The tracks in this oblique view up the rock-face, were left by a pair of titanosaurids, which were the largest of the herbivorous sauropods. They tramped across the soft lime mud that was almost a crust on the lake flats, squishing it into ridges around almost perfect footprints that are each 300–400 mm long. Subsequent tectonism created joints aligned down the dip, and also the en echelon sigmoidal tension gashes along a shear zone that almost parallels the tracks. Elsewhere, some footprints have punched through the carbonate into soft mud below, and others stand proud as pads of compressed mud left on top of the same carbonate. Dinosaur trackways have been found at multiple horizons within the quarry.

Fortunately, the cement factory cannot use the dolomitic carbonate that forms the trackways bed and those beneath it; so they are just stripping away the overlying pure limestones and clays, while the trackways will remain intact. A few of the footprints were first found during the 1980s, but the quarry rose to fame in 1994 when many more were revealed, including a single unbroken track 350 metres long. As extraction of the valuable limestone continues, additional lengths of the trackways bed are being exposed along the quarry wall, and yet more tracks are likely to be revealed.

Cerro Rico at Potosi

It means "Rich Hill", and it has probably been the richest hill anywhere in the world, based on the enormous mineral wealth that it has yielded over the centuries. At an altitude of more than 4000 metres, on the eastern fringe of the Bolivian Altiplano, it rises next to the city of Potosi, which owes its location purely to the mineralisation and the mines. First worked by local people in the 1500s, at the behest of the Spanish conquistadors, the mines have yielded well over 30,000 tonnes of pure silver from ores including primary sulphides and a secondary chloride. Subsequent phases of mining have produced vast quantities of tin, zinc, lead, copper, antimony, tungsten and other metals, which have been extracted from a variety of ore minerals.

Cerro Rico is the core of a Miocene volcano, which was once an active component of the Andean plate boundary. Nearly 14 million years ago, a dacite dome rose within the volcano and expanded to more than a kilometre across. Cooling fractures within the dacite were then permeated by ascending hydrothermal solutions that deposited minerals, producing a zoned profile throughout the volcanic edifice. Supergene enrichment, driven by descending meteoric waters, was a bonus that followed later, improving metal values in what are now the upper parts of most veins. However, the only significant surface workings have been the more recent ventures extracting silver and tin from extensive alluvial gravels known as pallaco, which are up to 70 metres thick on the lower flanks of the mountain. There has never been a large open-pit mine at Cerro Rico, Instead, hundreds of mineral veins, in numerous systems, have yielded rich ores to traditional underground mining techniques that have matured and improved slowly through the centuries. Even now not everywhere within the hill has been reached by the miners. The richest veins and largest mines are now gone, but to this day hundreds of miners continue working in small co-operatives. Each group has its own adit entrance and its own small stopes, where they drill shot-holes by hand, shift ore with shovels, hand-winch buckets up internal shafts, and shoulder tubs along adit railways. It seems that time has stood still inside Cerro Rico.

Situated within the Parque Nacional Lauca high in the northern tip of the Chilean Andes, Parinacota is a volcano with a snow-capped cone that rises to 6348 metres above sea level. It is a singularly beautiful composite volcano, but the main interest in its geology lies within an enormous debris avalanche, which is one of the finest to be seen anywhere in the world. The entire cone of the modern volcano, which rises nearly 1700 metres above the surrounding highlands, is post-glacial. Its predecessor, which was an even larger volcanic edifice on the same spot, collapsed in a single massive event about 8000 years ago. The western side of the original volcano failed completely in a catastrophic rotational landslide. This produced a massive debris avalanche that swept down and away to westwards (towards the camera, then off to the left of this view). More than six cubic kilometres of rock debris were spread out over 156 square kilometres of land, with a maximum run-out of 23 kilometres.

The head-scar of this giant landslide is buried inside the later volcanic cone, but the nature and extent of this debris are hugely impressive. Within it, hills up to 400 metres across and 80 metres high are single blocks of volcanic rock that came away from the original mountain. The rest is a chaos of lava and tephra that has travelled far from its original site. Finer tephra fragments have been winnowed by the wind and now form accumulations in sheltered hollows. The rock avalanche filled a valley and created dams that hold back a series of lakes, which continue to drain out unseen through the porous debris barriers. This style of lateral collapse of andesitic volcanoes has been fully recognised only since the 1980 event at Mt St Helens in Washington State, USA, even though that event was more notable for its lateral blast and less so for its debris avalanche. Prior to that, nobody witnessed the remarkably similar 1956 event at Bezymianny in Russia's Kamchatka. However, the combination of collapse and rock avalanche is now recognized as a major process in a great many volcanic terrains, and among these cases Parinacota provides a textbook example.

Chuquicamata copper mine

Out in the totally dry emptiness of Chile's Atacama Desert, Chuquicamata is an enormous hole in the ground. Its orebody is one of the huge 'porphyry copper' deposits that are characteristic of the convergent plate boundaries along the Andes and Rockies mountain chains. Chuquicamata's metal ores originated within multiple igneous intrusions and late-stage hydrothermal mineralization that developed during Palaeogene times about 40 million years ago. Chalcopyrite and bornite are the primary copper ores, though there is a thick cap of supergene enrichment to chalcocite, and molybdenite is a significant by-product.

Open-pit mining is the only method of working this type of orebody economically, and the pit is now 4500 metres long, 3500 metres across and an astonishing 800 metres deep. All the mine benches are blasted back in sequence and the broken rock is hauled out in trucks, each of which carries more than 300 tonnes of rock and takes 45 minutes to be driven up out of the pit. The targeted ore is worked from the lower benches that are within the roughly cylindrical orebody, while the upper benches have waste rock removed merely to maintain a stable pit profile. Daily output is half a million tonnes, of which just one third is ore, and that contains only 1.15% of metals. A massive plant on the site treats the ore by crushing, flotation and electrolysis to produce an annual 600,000 tonnes of pure copper (and 13,000 tonnes of molybdenum as a major by-product). Waste heaps spread across the desert for many kilometres in every direction.

The mine employs around 10,000 workers, many of whom are now being re-located (with their families) to the nearby town of Calama so that new waste heaps can bury the old company town and eliminate haul times to more distant tips. Chuquicamata is the world's largest pit, and was the world's largest copper producer, with statistics long vying with those for Utah's well-known Bingham Canyon Mine, though it now produces less copper than Chile's newer Escondida Mine. All three giant mines are in 'porphyry copper' orebodies.

Gruta do Janelão

Within the uplands of southeastern Brazil, the broad valley of the Rio São Francisco has a surprise for any passing geologist, in the form of the Peruaçu National Park. This encloses an area of limestone karst little more than 20 kilometres across, which contains some truly spectacular landforms. The strong and pure Precambrian limestone forms a low, forested plateau that is traversed by the Rio Peruaçu. Parts of this river's course are entrenched in canyons up to 150 metres deep, but the intervening sections pass through some exceptionally large cave passages. The longest underground section extends for more than three kilometres through the Gruta do Janelão. Not the largest known cave passage in the world, but right up there in the top ten, this is probably the easiest to be appreciated because so much of it can be seen by daylight. The cave has its roof broken by four giant skylights, as reflected in its name, which translates as the Cave of the Windows. Parts of the passage are around 100 metres high and wide, and through the entire three kilometres there is only one short stretch that does not have a glimmer of natural light from one of the windows or giant entrance porches. The cave river winds its way between banks of sand and gravel, but there are also some massive cascades of calcite terraces down parts of the walls. Furthermore, the outsides of the cave windows are giant sinkholes. These can be fully appreciated from viewpoints that break out of the forest, and are well worth the extra treks to reach them through this spectacular National Park.

Since this photograph was taken, the cave has been made more accessible for the small numbers of visitors who reach the rather remote location. With a local guide, they can traverse the giant cave passage, as long as they can cope with the 1300 stone steps that provide access into the depths. The new walkways, completed in 2014, are not obtrusive, but would be just visible in front of the person silhouetted on the far sandbank. Gruta do Janelão is rapidly becoming known as a must-visit location for adventurous travellers in South America.

Trinidad Asphalt Lake

Near the southwestern corner of the Caribbean island of Trinidad, the village of La Brea takes its name (in Spanish) from the adjacent tar pits that include the Trinidad Asphalt Lake. Natural asphalt is an impure form of bitumen, and may be known colloquially as tar. One of only three tar lakes in the world (the others being in Venezuela and California), this is formed by an active, natural seepage of hydrocarbons to the surface, where the volatile components are then lost to the atmosphere. Source rocks are Cretaceous marine mudstones far below, which feed into Tertiary sandstones that have become reservoir rocks for the many nearby oilfields. Some of the heavy-fraction tar rises along intersecting faults to form both the Asphalt Lake and the many other smaller tar seepages nearby.

The Lake extends over an area of about 40 hectares, most of which has a black rubbery surface that becomes notably softer when warmed by the noon-day sun. About a fifth of its area has a surface of warmer, more-liquid tar that has emerged anew from below. Rainwater can lie in shallow pools over both types of tar. There is also some limited upwelling of sulphurous water, along with methane gas that is sometimes seen as small dancing flames. Drilling to a depth of 60 metres into the lake-bed found the same asphalt all the way down. The exposed material is a valuable resource, which is still used to manufacture road-surfacing material. Annual extraction now runs to about 9000 tonnes; it is simply scraped out from trenches that are excavated to a metre or so deep with machine shovels. Then within a week or so these refill naturally with new tar that rises slowly from beneath. However, over the years a total of more than 82 million tonnes amounts to over-abstraction, so the Lake now lies inside its own crater that is about nine metres deep, whereas 200 years ago it was level with the adjacent ground. Away from the extraction machinery, the Lake now also provides a modest visitor attraction, and guides are on hand to ensure that over-inquisitive geologists don't take one-way trips into the areas of soft and warm tar.

Panama Canal

When it was completed in 1914, the Panama Canal was the world's greatest feat of civil engineering. Its construction still ranks as a hugely impressive project, especially since the addition of new giant locks that are capable of taking most of the world's larger container ships, which have long outgrown Panama's original locks. Lying close to each end of the canal the flights of locks take ships to 26 metres above sea level. Even at that altitude the canal is still deeply entrenched in the Culebra Cut, where it breaches the watershed hills. For passengers on the many cruise ships that now transit the canal, one of the most memorable sights is the massive, stepped, rock wall cut into Gold Hill, on the east side of the Culebra Cut. However, this is not the greatest feature of the canal. Nor is this the best photograph of Gold Hill, but it does show, on the left, just part of one of the enormous landslide scars that were created during excavation of the canal.

Following the French withdrawal from the project, American construction crews faced horrendous ground conditions where they had to cut the canal's route through the watershed hills. The geology consists of strong sandstones and notably weak clays, all with components of volcanic origin, and all highly weathered and heavily broken by faults. Explosives were needed to progress through the strong sandstone of Gold Hill, but the cutting was then stable in a steep wall. The major problem was with the weak Cucaracha Clays, which sheared so easily at the base of huge landslides. During the canal's excavation phase, these slides slumped into the new cuttings and were eventually removed in their entirety by continually shovelling away the debris toes while the cuttings grew ever wider. Well to the left of this view, the head scar on the East Culebra Slide lies nearly 400 metres back from the canal. The entire landslide was removed, to a depth of more than 100 metres. The great bowl left where the landslide once stood is easily overlooked by shipboard tourists gazing up at the rock face of Gold Hill, but dealing with the clay-based landslides was the real success story of engineering geology at the Panama Canal.

Banded amber

All Mexican amber comes from the country's southernmost state of Chiapas. It is currently produced only on a small scale that bears no comparison to the production from numerous mines in the Dominican Republic or the large opencast mines along the southern Baltic coast. A single basin of Late Oligocene shales and sandstones amid the mainly limestone mountain ranges north of San Cristobal de las Casas is the only source in Chiapas. Lumps of the amber had long been found within the alluvial sediments of the basin by the indigenous Mayan people, who were working the mineral into jewellery more than a thousand years ago. But Chiapas amber was 're-discovered' only in the late 1940s. Nowadays, most of it is extracted from a handful of small mines, with fairly grim underground working conditions, all dug into the hillsides around the small town of Simojovel. Even these form only a cottage industry, largely with seasonal working when most of the miners are not involved with the coffee crop that is the mainstay of the local economy.

The precious mineral comes out of the ground in the form of dark lumps, and the radiant colours become apparent only after a dirty crust has been chipped away with great care. Chiapas amber is notable for its wide range of colours, from the palest yellows through to deep reds. Like all amber, it is fossilized resin, and in this area is largely from trees in the *Hymanaea* genus, which include the same species that formed the Dominican amber and still grow in parts of South America. Banded amber, which is rare, appears to represent annual layers of resin that each flowed down over the previous season's deposit on the outside of the tree's bark. As is usual among ambers, numerous fragments of organic debris were caught in the resin, which was significantly sticky when it was fresh. This piece, which has been cut and polished to be made into a lenticular pendant 50 mm long, contains five small complete insects; their crumpled remains cannot readily be identified, but they are probably some type of fruit fly.

Not a pole, but a hollow steel tube, it's actually a well-casing, still in place but not in use. And it sticks out of the ground because the ground has subsided while the casing has remained static. This is the world's most extreme case of ground subsidence, and this most spectacular expression of the ground movement is in the Plaza de la Republica, one of the more important squares within central Mexico City. It is more than a little sad that most people pass by unaware of the story behind the preserved, black-painted 'pole'.

Mexico City is built on an old lake bed that is floored largely by horribly weak, highly compressible, smectite-rich clays. These cause settlement of many of the city's old buildings under their own weight, with some sinking well over a metre. In contrast, modern structures have appropriate foundations designed to eliminate virtually all ground movement that can be induced by their own loads. The major problem lies at depth, where sands interbedded with the clays have long been exploited for supplies of fresh water. Pumped abstraction of the groundwater has almost no impact on the sand aquifers themselves. However, water pressures within the alternating sands and clays ultimately equalize, as water is squeezed out of the clays and into the partially drained sands. That loss of support by the pore water pressure causes the clays to compress under the weight of the geological succession, independent of any imposed loads from buildings. The result is regional ground subsidence on a grand scale, and by as much as nine metres in the heart of Mexico City. A large part of the city centre has subsided by that amount within the last hundred years. The only structures not to have subsided are the well casings, which are effectively founded in the aquifer sands a hundred metres down and below most of the compacting clays. Controls on water abstraction have now reduced the rate of subsidence, but have not stopped it completely. The well-casing had been freshly painted for only about a year when a centimetre-wide band of unpainted metal was already exposed, where the ground had subsided around the unmoving steelwork.

The Copper Canyon ignimbrites

Famed as being deeper and larger than the Grand Canyon of Arizona, the Barrancas del Cobre (which translates as Copper Canyon) is cut into the Sierra Tarahumara, a dissected plateau standing about 2500 metres above sea level in northwestern Mexico. The Copper Canyon is more than a single feature; it is a complex, dendritic, drainage system with a trunk river flowing 1700 metres below the canyon rim. Visual impact of the canyon and its branches is enhanced by great lengths of their rims being formed within a sub-horizontal bed of strong, pale-coloured ignimbrite, which stands in near-vertical walls about 100 metres high atop the canyon's complex of cliffs and ramparts.

This rock is part of a sequence more than 1000 metres thick that covers an area 1200 km long by 300 km wide, and constitutes the world's largest silicic volcanic province. The enormous mass of ignimbrite overlies another major pile of volcanic rocks, also of Palaeogene age, that are dominated by andesitic lavas and pyroclastics. This older unit represents island arc volcanic rocks that erupted over a major subduction zone. Its phase of plate convergence was followed by an extensional regime that allowed the younger sequence of silica-rich pyroclastic flows to erupt through massive fissure systems. Activity lasted for about ten million years and the huge volume of ignimbrite created during those explosive eruptions suggests that there might have been more than 200 caldera vents.

The volcanic province also contains the world's largest known concentration of silver minerals, which were deposited at low temperature in epithermal veins within the lower volcanic rocks, at the time when the upper volcanic sequence was being erupted. During historical times, these veins were worked in numerous mines, most of which were along and close above the canyon floors, where the mineralized lower volcanic rocks were exposed. But that generation of rich mining in the canyon was all over by the early 1900s. The name of the Copper Canyon derived from the first Spanish prospectors into the area who initially thought they had found outcrops of green copper mineral. Though these turned out to be lichen, the minerals were there, and discovery of the silver riches followed.

Lava stalagmites in Apua Cave

Because Hawaii lies over a mid-Pacific hot-spot, its islands are formed almost entirely of basaltic lavas. Their characteristic structure is a pile of long lava flows that poured down the flanks of the large shield volcanoes, and most of those lava flows are tube-fed. Except where volumetric flow rates are high enough to maintain their temperature for some time or some distance as exposed lava streams, lava flows cool on their surfaces and thereby develop solid crusts. With further heat loss almost eliminated by the presence of that crust, flow of molten material can continue underneath, through a lava tube. One such tube, now known as Apua Cave, was formed in lava pouring down the southern flank of Kilauea during the 1973 eruption from the Mauna Ulu vent. Originally, Apua's tube was unseen and unknown, until a part of its thin roof collapsed during a small earthquake in 1975. This collapse allowed access to the cave, and revealed a remarkable underground scene.

In the closing stages of the 1973 eruption, the declining rate of lava production had allowed the Apua tube to drain down so that its floor solidified along an open tunnel. With no roof holes to allow escape, superheated gases continued to blast down the tube, and were hot enough to re-melt a surface layer on parts of the roof to create a glaze of basaltic glass. With continued heating by the gases, so much of this re-melt was formed that it dripped from the roof, with some of it then solidifying into thin lava stalactites. Yet more of it dripped to the floor, where it accumulated to form lava stalagmites. Sometimes known as lavamites, these grew taller as the succession of lava blobs landed on their tops, with each blob congealing where it landed and contributing to the distinctive knobbly profile of the stalagmite. Lava stalactites and stalagmites are known in various lava tubes around the world, but the gases appear to have been hotter or longer-lasting in Apua Cave. So many of the stalagmites grew to well over a metre tall, in a display of geological artistry that is truly exceptional.

How close can you get to a lava flow?

'Quite close' is the easy answer, as long as it is a peaceful, fluid, basaltic lava. Whereas the answer would be markedly different for viscous andesitic or rhyolitic lavas in explosive eruptions, basaltic lavas can be relatively people-friendly. Although a lava-stream within an established channel can flow as quickly as water, the front of a lava flow rarely advances at a speed that even approaches walking pace. So a basaltic lava flow in action can be a delight for visitors and tourists. However, the same flow may be less than welcome for those whose houses are slowly over-run by lava that is, all too often, effectively unstoppable. Hastily-built earth barriers to divert a lava flow, or sprayed water to cool it enough to prevent further movement, are options available only where valleys or water supplies are conveniently located.

The scene in the upper picture is on the black-sand beach at Kalapana, on Hawaii, when lava was reaching it from the Pu'u O'o vent on the Kilauea volcano (though unfortunately after having destroyed much of Kalapana village). This was a classic pahoehoe flow, with its front and sides spreading slowly by successive lava toes breaking out from beneath the cooler crust. It was readily accessible lava, but the person's approach was made more comfortable by a stiff, cool breeze blowing towards the lava; the radiating heat from the mass of hot rock was considerable, and also caused the shimmer that softened the photograph. During the next stage of cooling, the crust grows thicker, and it is possible to walk over the hot ground, even when red lava is visible down cracks just 200 mm below the surface. Perhaps the fluidity of basaltic lava can best be demonstrated by sticking a spoon into it.

The pictured spoonful of lava was scooped from the edge of an aa flow on Sicily's Mount Etna volcano (where the intense radiating heat did necessitate use of a spoon with a long handle). The lava dripped off the edge of the spoon for a minute or so before it cooled sufficiently to become immobile; and ten minutes later it was all cool enough to tuck into a rucsac.

Generally acclaimed as the most beautifully decorated cave in the world, Lechuguilla is also one of a particularly special type. It lies in the Guadalupe Mountains, very close to the well-known Carlsbad Caverns, in New Mexico, USA. The cave was discovered in 1986 when cavers engineered their way through a draughting boulder choke at the foot of a 30-metre-deep shaft isolated on a remote desert hillside. But this single entrance now gives access to more than 220 kilometres of mapped cave passages.

Lechuguilla Cave contains no underground streams, because it was not formed by sinking rainwater, which is the most widespread process of dissolutional cave development. Instead it was formed from below, by fluids that migrated upwards into the great mass of Permian reef limestone fringing the huge Delaware Basin. These fluids were rich in hydrogen sulphide that was produced by degassing of hydrocarbons in contact with basinal gypsum beds. Similar hydrocarbons are now pumped from the oilfields. Where the hydrogen sulphide rose into the marginal reefs, it was oxidized by descending,

oxygen-rich meteoric groundwater to create sulphuric acid. This acid was extremely efficient at dissolving the limestone, and it produced huge amounts of new gypsum as part of the process. In this way, the caves of Lechuguilla and Carlsbad (and some others at higher elevations) formed over a period of about 12 million years, mostly at and around a steadily falling water table within the evolving limestone mountains. Crusts of pure white gypsum are widespread in Lechuguilla, but one of the cave's many highlights is the collection of gypsum chandeliers in the wide chamber known as the Chandelier Ballroom. The entire cave system is now almost dry, but condensation within chambers about 25 metres above the Ballroom produces water that seeps down through massive floor deposits of gypsum. This acidic seepage evaporates where it emerges through the Ballroom ceiling, to deposit the huge branching crystal masses that constitute the chandeliers. These hang down by as much as six metres to create a rather supernatural, but truly fantastic, underground scene.

Monument Valley

Lying largely within Arizona and Utah, America's great Colorado Plateau is commonly known as Red Rock Country because of its dramatic desert landscapes that are dominated by Permian and Triassic red sandstones. Close to the plateau's centre, and just inside Arizona, Monument Valley has long been a homeland for the Navajo tribe. Then since 1939 it has become an icon. both for the region and for Hollywood Westerns, after the film director John Ford used it as a backdrop for John Wayne's exploits in *Stagecoach* and also in nine subsequent films. The Valley's narrow buttes and wider mesas all owe their existence to erosion and dissection of the edge of a sandstone plateau, where the almost horizontal beds include a thick unit of strong sandstone that has been undercut by weathering and erosion of a relatively weak sequence of shales and thin sandstones.

In and around Monument Valley, all these spectacular landforms have near-vertical walls about 120 metres tall that are formed in the massive, red, aeolian De Chelly Sandstone. These great rock walls rise above stepped plinths that have developed in a similar thickness of interbedded shales and sandstones of the deltaic Organ Rock sequence. Another strong bed, the Cedar Mesa Sandstone, forms the relatively level floor in the lower part of the valley between the Merrick and Mitten buttes that are seen in this famous view from Artist Point.

Even though the site now lies within a semi-arid desert, the great majority of the erosion that has left the buttes behind has been fluvial, and wind erosion has contributed little. Streams drained off the edge of the main sandstone plateau to cut deep canyons by headward erosion. Their sides then retreated, not by the usual outward flaring and degradation, but by the spalling of slabs of rock between vertical joints that opened progressively owing to lateral stress relief in the ground alongside the valleys. Only when the valleys had cut deep enough to expose the weaker Organ Rock beds, did they develop with V-shapes at those lower levels. There the valley floors expanded and undermined the great sandstone walls, which have then maintained their vertical profiles by continuing to retreat on successive vertical fractures.

Rainbow Bridge

Located almost at the centre of the magnificent red-rock deserts of America's Colorado Plateau, Rainbow Bridge is a textbook landform. It stands in a deeply entrenched canyon that is a tributary to the Colorado River in southern Utah. Surrounded by wilderness, few people other than itinerant Navajos saw it until it was 'discovered' by a team from Utah University in 1909. The long hike in has only been avoidable since the 1960s, when construction of the Glen Canyon dam impounded Lake Powell, and its waters backed up to just beneath the bridge. Now it can be reached easily from Page, in Arizona, by a glorious boat trip (which no passing geologist should miss). The approach to the bridge is along a half-flooded, meandering canyon, and the first view is magnificent, either from the boat if the reservoir level is high, or from a short footpath when the waters are low.

Rainbow Bridge is massive. It has a clear span of 85 metres and is 88 metres high; its top is 13 metres thick and 10 metres wide; it is one of the largest natural bridges in the world. It is a bridge, and not an arch. Bridges are undercut by rivers that break through the necks of deeply incised meanders. In total contrast, but with a confusingly similar end-result, arches are created by weathering that causes face retreat on both sides of a thin, joint-defined, blade of rock known as a fin. Rainbow Bridge is all that is left of a single tight meander core where two bends of the river converged at floor-level within their canyon. Extrapolation of erosion rates suggests that Bridge Creek started to incise its meander around the bridge site about 125,000 years ago, and has been flowing under the bridge for about 30,000 years, leaving its dry meander channel wrapped around the bridge's eastern footing. The Jurassic Navajo Sandstone has few joints or major bedding planes, so the bridge's rock is almost flawless, which is why it has survived high above its stream, and neither erosion nor weathering will see its demise for thousands of years to come.

Antelope Canyon

The Red Rock country of America's Colorado Plateau is symbolized by towering vertical cliffs and deep twisting canyons. Among the multitude of highly photogenic sites, Antelope Canyon is one of the more beautiful and more accessible. It is cut into the Jurassic Navajo Sandstone, where the textures of the red dune-bedded sandstone make any outcrop a delight to behold. Lying just east of the town of Page, close to the Arizona–Utah border, Antelope Creek drains a wide sandstone ramp that slopes gently down to the Colorado River where it is now impounded in Lake Powell. Along with the other creeks on this great sandstone ramp, the course of Antelope Creek alternates between sections with wide sand floors and others that are narrow slot canyons. Occasionally, all are swept by powerful flash floods.

Originally a small tributary of Antelope Creek flowed across a higher section of the gently undulating sandstone ramp, so that effectively it was perched, by about 30 metres, above the main creek. It has therefore entrenched itself into the sandstone, by cutting a narrow canyon headwards from the confluence. Nearly all of its erosion has been by flash floods that are generated by localized storms over the mountains some kilometres away. There is almost no local surface run-off that can erode its flanks, so the canyon has developed into a classic slot gorge, with vertical and overhanging walls that wrap round the incised curves.

Known as Antelope Canyon, even though it carries only a tributary of the main creek, its easily accessible section is little more than 30 metres deep, but it is only a few metres wide, so the sky is mostly lost from sight above the twisting walls. The canyon's tight curves evolved into different shapes while cutting ever deeper into the sandstone; they are not meanders in the classic sense, because they have been scoured by floodwaters and there is no room for sedimentation inside the bends. Though only about 800 metres long, the slot canyon of Antelope Creek's beautiful tributary is a classic of fluvial geomorphology, where the finest products of human architecture are thoroughly eclipsed by the wonders of natural erosion.

Angels Landing, Zion Canyon

Among the multitude of inspiring landscapes across America's Colorado Plateau, Zion Canyon, in southwestern Utah, has to be one of the highlights. Cut into 600 metres of remarkably massive, dune-bedded, red and buff, nearly horizontal, Jurassic, Navajo Sandstone, Zion is characterized by its towering cliffs and vertiginous views. Midway along the canyon, Angels Landing is the whimsical name for the flat top of an almost isolated rib of sandstone that stands 400 metres above the canyon floor. It is readily accessible by way of a well-engineered National Park trail, which offers one of the world's most exhilarating short walks for those of fairly energetic disposition. The trail climbs into the joint-guided Refrigerator Canyon (which fortunately lies in shade early in the morning) and then zig-zags up its wall to a saddle at the neck of Angels Landing (from where this photograph was taken). A vertical wall drops 300 metres back to the main canyon floor, with bird's-eye views into the Narrows.

The trail continues along a narrow ridge, with lengths of chain to act as handrails along the exposed sections. On one side lies a spectacular vertical drop to the floor of the main canyon; on the other a rock slope almost as steep falls away into the tributary canyon. Then the trail rises yet further onto a slightly wider summit that offers glorious views in all directions. Downstream, Zion Canyon is a massive feature whose near-vertical walls have retreated by undercutting of a weaker sandstone unit that is exposed near floor level. In contrast, the upstream Narrows have not cut down as far as this weaker sandstone, so retreat of its nearly vertical walls has been negligible, and the canyon is a simple slot-gorge that is little wider than its river bed. Many of the landform details are guided by joints within the sandstone, though the Angels Landing also gives a bird's-eye view of a magnificent incised meander where the main canyon loops away before joining its Refrigerator tributary. The walk (or perhaps it should be described as a scramble) back down the upper trail gives plenty of time to appreciate the precipitous sandstone landscape: this site has geology and scenery on a dramatic scale.

Dead Horse Point

A bare-rock landscape of cliffs and benches provides one of the world's great geological vistas. America's Colorado River has cut down through 600 metres of almost horizontal red rocks in the semi-desert wilderness of what is now known as Utah's Canyonlands National Park. Although the red-rock country of the Colorado Plateau is well known, many visitors miss this single best viewpoint, tucked away in a small Utah State Park at the end of a spur road some 50 kilometres from Moab. The view is not of Dead Horse Point, but is from it. Far below, bright red Permian sandstones rise from the river to a cap of pale Shafer Limestone that forms the bench within the incised meander. Cliffs of red Cutler Sandstone continue up to the White Rim Sandstone (which thickens southwards into the De Chelly Sandstone well-known where it forms the Monument Valley mesas). The distant high red cliffs beyond the White Rim bench are the Triassic Wingate Sandstone, capped by the paler Triassic–Jurassic Navajo Sandstone; these two form most of the big red cliffs in Utah and Arizona. The slopes beneath the Wingate are in red Moenkopi siltstones and green Chinle shales, which contain the Petrified Forest farther south and some marginal uranium resources in this area.

Before the site became a national park, prospecting for uranium was encouraged by the state government, which financed construction of the dirt road on the bench at the foot of the photograph. Uranium was never found in extractable quantities, but the long and rough track became popular with off-road car enthusiasts, and then gained cinematic fame as the final destination for Thelma and Louise. From far above that bench, the view from Dead Horse Point brings geological time alive: both the hundreds of millions of years taken to form the great sequence of sedimentary rocks, and then the millions of years for the Colorado River to erode into them and shape this huge, sprawling canyon. This is geology at its best.

Crystal Geyser

Geysers are well known as the more spectacular highlights of geothermal areas. Beside the original Geysir, in Iceland, after which they are all named, but is no longer active, there are dramatic erupting geysers in Yellowstone, Rotorua, Kamchatka and Iceland, but there are few of any size elsewhere. All of those consist of boiling water blasted skywards by steam. So a cold geyser is an oddity. Crystal Geyser, in Utah, is one of the few, and is probably the world's largest. It lies at a remote spot beside the Green River, at the end of eight kilometres of dirt road leading into the semi-desert south of the town of Green River.

The vent is an abandoned oil well, but its eruptions are naturally driven. They occur only once or twice a day, and each starts with water rising within its vent so that it overflows for about 20 minutes. It then suddenly erupts to its full height of more than 20 metres, until ceasing all activity after about another five minutes. It is powered by carbon dioxide, and the source of both the gas and the water is the porous Navajo Sandstone at a depth of about 215 metres. The carbon dioxide is probably generated by acidic groundwater from deeper beds of gypsum reacting with carbonate cement in the sequence of overlying rocks. The gas goes into solution in the groundwater until saturation is reached, when it exsolves and causes the water column within the old well to effervesce. This then pushes water out of the well casing, so that the reducing pressure from the shortening head of water allows yet more gas to be exsolved from groundwater within a limited area of influence of the well. Eventually, rather as in a steam-driven geyser, a chain reaction accelerates into a flash eruption, which finishes when the well-casing is emptied. The pressure loss and consequent exsolution are localized by the low transmissivity of the aquifer, and the sudden expulsion of all the gassy water allows the well to refill with groundwater from the surrounding aquifer. This influx of water has remained under-saturated with respect to carbon dioxide, so stability is restored until further gas production deep underground starts the next eruption cycle. The bonus is that, without steam to obscure the view, Crystal Geyser is delightfully photogenic.

Exfoliation on Half Dome

Yosemite National Park, deep in the heart of California's Sierra Nevada, is everything that a national park should be. Spectacular mountain scenery attracts hordes of visitors, including significant numbers who go to climb on the great walls of clean granite. Geologists too are drawn to the many splendid features of the granites and granodiorites within the multi-phase batholith that is exposed so well across the glaciated terrain. Among the more spectacular features of the main granodiorite are its huge exfoliation domes, typically a kilometre or so across and some hundreds of metres tall. These domes are essentially features of stress relief, with their huge roughly concentric fractures developing near the surface as cover rock was stripped away by erosion. Stress distribution within the ground means that successive domed fractures, each developing inside its predecessor, tend towards spherical shapes between the deeply entrenched valleys. Subsequent erosion then leaves the massive domes with shells of rock, each a metre or so thick, that fall away progressively as each mountain evolves towards an ever-more-rounded dome. Cooling fractures had previously developed within the batholith, but these alone cannot account for the shapes and patterns of the beautifully domed mountains.

Half Dome, rising far above Yosemite's main valley, is only a half because its near-vertical face along the valley wall is defined by a major set of tectonic fractures that are older than, and totally independent of, the domed exfoliation fractures. The face was trimmed clean by Quaternary glaciers, but there never was another half of a dome that was entirely removed by the ice. Away from that wall, the top, back and sides of Half Dome have all the features of a classic dome on a truly grand scale. This view looks steeply up its northeastern side, with the main face in oblique profile on the right. The exfoliation shells are perfectly revealed, and scale is given by the fourteen hikers making their way up or down the shared cable-ladder on the far left. The long and hard hike up Half Dome, all the way from the valley floor and then up the cable-ladder, is one of California's great outdoor experiences, but it also gives the energetic geologist a rare chance to get up-close and personal with exfoliation fractures at their finest.

Mount Rushmore in South Dakota, USA, is well known for its giant sculptures of four heads of past American presidents hewn into a rock wall. Those carvings were completed in 1939, whereas work on creating the Crazy Horse Memorial is still in progress on the nearby Thunderhead Mountain. This is a huge rock carving of the great Lakota leader and warrior who was active in the late 1800s. Each of the Mount Rushmore heads is 18 metres tall, whereas the head of Crazy Horse is nearly 28 metres high; and he is mounted on a horse, so that the final carving will be 171 metres tall. This incredible project was initiated by the Lakota people when they called in the Polish sculptor, Korczak Ziolkowski, who started work in 1948. It is funded entirely by donations and visitor fees, which is why the head was completed to be a visitor attraction ahead of progress on shaping the rest of the mountain. It will probably be another 60 years before the carving is completed, under the direction of Ziolkowski's descendants.

Both mountains, of Thunderhead and Rushmore, are formed in Proterozoic granitic rocks that form the core of the breached dome of the Black Hills of Dakota; the black in their title derives from the dark cover of coniferous forest that stands out from the surrounding, paler, grasslands. Whereas Rushmore is formed of a granitic gneiss with bands of migmatite and schist clearly visible across the presidential faces, Thunderhead (13 kilometres away to the west) is in a batholith of mature, porphyritic granite with little internal structure. Just below the left eye of the carving, a flat cleavage surface broken through a large feldspar crystal can sometimes catch the sunlight, at which times it is known as Crazy Horse's teardrop. Machinery visible on the skyline in this view stand on what will eventually become Crazy Horse's out-stretched arm. This will reach as far as his horse's head across a void that will be opened out from the tunnel already in place with its portal in shadow below the machinery. This will be a truly amazing granite sculpture.

Niagara Falls

Perhaps the best known, certainly not the highest, but truly impressive, Niagara Falls may be famous along the American tourist trail, but they have also given their name to a specific type of waterfall. Their clean drop of 55 metres is perpetuated by the strong Lockport Limestone forming a sharp lip that is undermined by expansion of the plunge pool in an underlying sequence of relatively weak shales. This combination of strong caprock over a weaker rock is what defines a Niagara-type waterfall. Wherever the geology is right, this is always a clean vertical drop, with easy excavation of a plunge pool in the weaker rock preventing any tendency for the waterfall to degrade into a ramp of broken cascades. At Niagara itself, the strong Lockport caprock (it's actually a dolomitic limestone of Silurian age) forms only half the height of the waterfall. Beneath it, the shales (with some interbedded sandstones) are exposed for about 25 metres and continue down for another 50 metres to the floor of the plunge pool.

The pairing of strong and weak rocks ensures that the waterfall retreats slowly upstream as chunks of the lip break off when they are undermined. This retreat is a characteristic of the Niagara-type waterfall, and is seen particularly clearly at the Horseshoe Falls (the Canadian half of Niagara) with its centre retreating faster because it carries the greater flow. Over time, the retreat leaves a steep-sided downstream gorge. At Niagara, the gorge is now more than 10 km long, cut back into the Niagara Escarpment where the waterfall originated. That was almost 11,000 years ago, when the last Quaternary glaciers melted away to leave the Great Lakes spread across the North American lowlands. The Niagara River was part of a new drainage system, carrying the water from Lake Erie, over the Niagara Escarpment and down into Lake Ontario. So the Niagara Falls retreat rate is about a metre per year, and that will continue until the Falls are eventually lost where the limestone dips beneath the waters of Lake Erie.

Morning Glory, Fading Glory

One of the best-known and most recognizable hot springs anywhere in the world is the lovely Morning Glory Pool, which lies within the Upper Geyser Basin of the extraordinary geothermal area encompassed by America's Yellowstone National Park. This colourful pool, just seven metres across and the same deep, produces a steady flow of water at well below boiling point. So it does not normally erupt into a geyser. It was seen to erupt just once, back in 1944, for uncertain reasons that were probably connected to a local geothermal anomaly. The pool gained its name because it was originally a beautiful, deep-blue colour, right to its edges, therefore looking rather like the Morning Glory variety of the *Convolvulus* flower.

Sadly, the colours have changed with the years, so that it is now also known, although not officially, as Fading Glory. This change has occurred because the spring has cooled down. Whereas some natural changes in the geothermal environment are ever likely to occur, most of this cooling has been due to the spring's reduced flow. The tragedy is that this flow reduction has largely been due to the spring's throat becoming partially choked by junk, most of which consists of low-value coins that have been thrown in by unthinking visitors. The pool's colours are a feature of the various thermophile bacteria that each thrive in water of a certain temperature. Critically, the reduced flow, and therefore the more rapid cooling on meeting the open air, has allowed low-temperature yellow bacteria to flourish in the water's upper layers. This photograph dates from the late 1970s when the pool's blue centre had gained only a yellow rim. The original road through Yellowstone passed right by the pool, until it was relocated in 1970, but tourists still walk to the site and still throw in their coins. So the changes continue, and the yellow bacteria are spreading downwards, turning the original blue into a rather less exotic green. It is an unfortunate fact that some of the more fragile features of the natural world are simply too sensitive to mankind's thoughtless intervention.

Straddling the crest of the Canadian Rockies, the Columbia Icefield feeds outlet glaciers that drain their meltwaters to the Pacific, Atlantic and Arctic oceans. Besides being the largest icefield in the Rocky Mountains, the Columbia is also remarkable in that a large part of it rests on limestone bedrock; and this contains the Castleguard Cave. The only entrance to the cave lies in a wooded slope well below the ice margin, but its single main passage reaches back into the mountain and then for three kilometres beneath and beyond the margin of the overlying ice cover. Following the bedding up the gentle dip of the Cambrian limestone, the cave passage eventually meets the floor of the Columbia Icefield. Its end, as far as human visitors are concerned, is this ice plug with a spectacular and beautiful wall that sparkles in the light of a caver's lamp. At this point the windswept surface of the Icefield is about 285 metres directly above. The level of the rock floor beneath the ice is uncertain, but is probably less than 50 metres above the open cave. From the hole in that rock floor, ice has been squeezed like toothpaste into and along the tubular cave passage. However, the few cavers who have reached the ice wall have recorded it as being practically stationary throughout the last 40 years.

Prior to the Quaternary glaciations, the icefield site was a wide karst basin, part of which drained out through Castleguard Cave. Its present situation demonstrates how ice can seal a cave passage and inhibit further karst development. In contrast to that, large flows of meltwater do drain into and through other parts of the cave during the summers. Castleguard is therefore a valuable indicator of how caves, in Britain and elsewhere, were reduced to minimal activity when they were covered by Quaternary ice that either blocked them or fed in meltwater with limited dissolutional capability. However, this unique ice plug is six kilometres from the cave's entrance, and much of that passage is difficult to traverse, so it needs a three-day underground trip to visit the site. The end of Castleguard Cave is a classic among sites of geological interest, but it remains one of the least visited.

Mendenhall Glacier

One of the world's most readily accessible glaciers is the Mendenhall. Some 20 kilometres long, this is an outlet from the great Juneau Icefield in southeastern Alaska, and it ends almost at sea-level, where it nudges up against the outer suburbs of the state capital, Juneau. The main bus route through town has one terminus at the glacier visitor centre, right on the Neoglacial terminal moraine that impounds the Mendenhall's pro-glacial lake. Across the lake, the ice wall at the terminus of the glacier is less than 20 metres tall, though it appears rather small as it now lies more than two kilometres away. Retreat of the ice terminus is currently around 30 metres per year, but the steady thinning of the glacier led to a major collapse in 2004, when large chunks at its front were left floating and then broke away to form icebergs in the lake.

Nowadays a close view of the glacier is best gained by hiking the excellent trail up the western side, which rises to a rock ridge that is 300 metres above and nearly two kilometres up-glacier from the lake. Right beside the trail over that ridge, the Mendenhall Glacier slides over its own buried rock step, so that the enforced curvature breaks its upper part into numerous crevasses. A mean rate of movement for the centre of the glacier is around 100 metres per year, and is probably about half that rate on the margins. That differential accounts for the development of another suite of crevasses that reach obliquely across the glacier. With the two sets of crevasses intersecting, the glacier is broken into a spectacular forest of seracs, many of which display a blue colour where sunlight passes through them. The overall effect is spectacular, and the view from the overlook is different every year. It is a splendid site of glacier activity, rather in contrast with a spot near the Mendenhall's eastern margin where a large flat expanse of ice is favoured for landing tourist helicopters when the cruise ships are in Juneau. This magnificent glacier has so much to offer, but the west-side trail should be a highlight of anyone's visit to Alaska.

Mendenhall glacier cave

The beautiful blue colour of so much glacier ice is simply due to the absorption of the red and yellow parts of the spectrum as light travels through ice. Generally, the surface of a glacier appears white due to light reflecting from the fractured surface layers of the ice, or from its cover of firn and snow. Light that has been transmitted through pure ice is best seen in deep crevasses or in glacier caves, but also in fracture faces that are sheltered from direct sunlight. And the blue is deeper and richer in ice that has been well compressed, so that it contains fewer of the air bubbles and crystal faces that can reflect much of the light. Generally, glaciers from the thick ice caps of the polar regions display more intense blue colour than do the smaller Alpine glaciers, which comprise less-dense ice. This is because the fewer diversions within the dense polar ice mean that light can travel farther through it, losing most of its red and yellow components along the way. So a cave inside an Arctic glacier has the best of the blue.

This example is in the Mendenhall Glacier, which is one of the outlets from the Juneau Icefield, among the mountains of southeastern Alaska. Glacier caves can be dangerously unstable, because ice can collapse at any moment in such an active environment; crevassed ice, seracs and hanging blocks can crash down with absolutely no warning. This was just a small cave beneath an unbroken wall of very solid ice at the glacier margin, so its floor was bedrock that sloped steeply towards the centre of the glacier. And, fortuitously, it was remarkably devoid of any morainic debris within the ice. So a mere few metres inside, blue light poured through a wall of clean ice that had melted away from the bedrock. The exposed surface of the ice had been carved into large scallops by ablation of water vapour into a steady wind blowing through the cave. The scene was magical; this was nature at its most beautiful. But it was also ephemeral. By next summer, the glacier had deformed and moved on, and its cave was no more.

Novarupta

In June 1912, Katmai volcano produced a huge ignimbrite flow and had its summit collapse into a caldera, on a scale that is unmatched within recorded history. These cataclysmic events were witnessed by no-one, and indeed nobody could have survived them at close quarters. Perhaps fortunately, Katmai is located in the remote and sparsely inhabited wilderness of the Alaska Peninsula, 450 km southwest of Anchorage. A tall ash-cloud was almost incidental to the huge 1912 eruption, and this was seen from afar. It covered the few coastal villages, 20 km away, with up to three metres of airfall ash, and also dropped 30 cm of ash onto the town of Kodiak, 160 km away. The eruption had three main bursts of activity, and lasted for a total of about 60 hours.

Nobody visited the volcano until 1916, when the Griggs Expedition, from Harvard University, sailed along the coast and then walked in to Katmai. They found the yawning caldera that had replaced Katmai's summit, and assumed that this had been the source of the 15 cubic kilometres of pumice-rich ignimbrite, most of which filled the valley of the Ukak River. This drains the western side of the volcano, and is now known as the Valley of Ten Thousand Smokes. Griggs also found the obviously fresh and aptly named Novarupta, but thought that it was merely a minor vent. Not until 40 years later did careful mapping reveal that this was the source of the huge ignimbrite flows that filled the Valley of Ten Thousand Smokes. A shared magma chamber allowed Katmai to collapse when the magma frothed out of Novarupta, some ten kilometres away. Since its formative eruption, Novarupta has been plugged by glassy rhyolite in the shape of a lava dome nearly 400 metres across. This sits inside a small asymmetrical ring of late-stage tephra, and both are surrounded by fractures defining their own miniature caldera. In this view, the high volcanic cone is Mount Griggs, an entirely separate volcano that has not erupted in nearly 4000 years. The skyline on the extreme right includes part of the rim of the Katmai caldera. Its jagged profile is all that remains of the tall volcanic cone that was there until that eventful day in 1912.

Valley of Ten Thousand Smokes

It no longer smokes, but it remains a dramatic volcanic wilderness, almost devoid of trees, shrubs or even grass. Alaska's Valley of Ten Thousand Smokes is tucked away amid the forests and volcanoes of the great peninsula, some 450 kilometres southwest of Anchorage. The entire valley floor is formed on the top of a massive volcanic ash-flow. This floor has minimal local relief except where it has been dissected by post-eruption streams, which now flow through sheer-walled ravines about 20 metres deep. The ash-flow, rhyolitic to andesitic in composition, was erupted as a dense mass of fluidized pumice. This swept slowly down the valley, without the energy and speed typical of a more mobile nuée ardente, or glowing cloud. There are about 12 cubic kilometres of pumice, filling the original valley to depths of up to 200 metres. At its upper end, near the vent, the ash-flow was hot enough to weld itself into an ignimbrite, but farther down the valley much of it remains more friable where it was only partially welded. This entire mass of debris exploded from the Katmai volcano during a cataclysmic eruption that lasted about 60 hours, in June 1912. Katmai's summit collapsed in a deep new caldera while the pyroclastic material was ejected through a new vent, now known as Novarupta, that lies low on the volcano's western flank. Emerging in the style of an overflowing champagne bottle, most of this dense volcanic froth took the downhill route, westwards, following the Ukak River and filling its valley along a length of more than 15 kilometres.

After the eruption, nobody visited the valley until a geological team led by Robert Griggs walked in from the coast in 1916. They found thousands of fumaroles producing plumes of steam all across the valley floor. Thinking that they had discovered a new version of Yellowstone, they gave the valley its new name. However, the steam was merely being created by remnant heat within the lower layers of the ignimbrite, which was boiling groundwater in the underlying alluvium along with any rainwater that reached it. By the late 1930s, nearly all the fumaroles had died away. Overlooked by the Griggs volcano (Katmai is barely visible from the valley floor), the valley is now silent, but it remains a scene of awe-inspiring volcanic devastation.

Gilkey Glacier

One of the major outlets from the Juneau Icefield, among the mountains of southeastern Alaska, the Gilkey Glacier is one of many that now end in their own pro-glacial lakes. This lake below the Gilkey has formed only since 1948, when the glacier's steady retreat saw its snout melt back from a low ridge of moraine. Since then the thinning lower end of the glacier has been breaking up into large and spectacular icebergs, which float away on a lake that is larger and deeper than many of its pro-glacial cousins. The icebergs gradually drift apart and diminish in size as they migrate down the length of the lake and eventually melt away. In common with most of the world's glaciers, the Gilkey continues to decline in size so that its face is retreating steadily, as it has done for most of the time since the end of the Little Ice Age around 300 years ago. Each year the ice-front is a little farther up the valley, and a new set of icebergs floats within the lake.

This part of Alaska came to be a little better known during the 1890s, when thousands of men came to the region, intent on crossing the mountains to reach the great interior valley of the Yukon River. These were the stampeders, who joined the gold rush that was world's greatest and craziest of all time. They were heading for the riches of the Klondike valley, a distant tributary of Yukon. And they had little or no concept of the hardships that they would encounter in these sub-Arctic mountains, which few people had even seen beforehand. Driven by the lure of gold, and guided by pathetically little information, they took various routes over mountain passes, across glaciers and through gorges, though it is likely that none encountered the Gilkey Glacier. At that time, it would have looked different, and was probably without a pro-glacial lake. It mattered not, as those long-suffering stampeders were in no mood to appreciate and enjoy the splendour and beauty of these glaciated mountains.

Moraines of the Meade Glacier

There is something rather beautiful in the sweeping curves of parallel medial moraines on a valley glacier that is fed by multiple tributaries. Along its path down from the Juneau Icefield, in the Coast Mountains of southeastern Alaska, the Meade Glacier is no exception. Almost dead centre in this view, a thin medial moraine is being created by the convergence of two lateral moraines where the Meade Glacier (entering from the right) is joined by its major tributary flowing in from the left. The two lateral moraines are barely visible, but their combined rock debris is drawn out into the dark and narrow medial moraine aligned towards the camera. This is all a result of the glacier's movement that is about half a metre per day. The broader medial moraine, left of the new one, derives from the convergence of two major glaciers pouring from the icefield far off to the left of this view.

On both sides of the conspicuous medial stripes, the many parallel curves of grey barely warrant description as moraines, but each band of dirty ice marks the input of a small tributary glacier from the adjacent rock slopes. Lower left in this view, an active lateral moraine is growing by accumulation of frost-shattered rock debris falling from the bare slopes immediately above the ice. However, this will never develop into a medial moraine, because there are no additional tributary glaciers before the Meade's terminus is reached in a pro-glacial lake another five kilometres down the valley. The zone of bare rock just above the glacier is conspicuous on the far walls, where a notably clean trim-line marks the highest level reached by the ice a few hundred years ago; this was during the Neoglacial advance, commonly known as the Little Ice Age. Above the trim-line, spruce trees thrive on thin soils that have formed and accumulated by weathering since the Last Glaciation, when the entire area was buried beneath a greatly expanded ice cap. Under the present climatic regime of slow global warming, the Meade Glacier is steadily reducing in size, but it should keep enough tributaries to maintain its beautiful medial moraines for many years to come.

Stone polygons

Peripheral to the glaciers, and hence extremely cold but not covered by ice, the periglacial environment is notable for the extent and variety of its patterned ground. These patterns have all been fashioned by repeated freeze and thaw of the soils within the few metres of the active layer that thaws each summer above the permafrost, which is the permanently frozen ground that lies beneath. Solifluction lobes are formed by slope movement, and ice-wedge polygons are the product of ground shrinkage, but the origins of stone circles and polygons within sorted soils are rather more enigmatic. This splendid example of stone polygons lies right beside the Denali Highway, just west of Tangle Lakes, in central Alaska. The site is on the floor of a small kettle hole, so the original soil is a classic glacial till of totally unsorted sediment (which nowadays is called a diamict and was formerly referred to as boulder clay). But this soil has now been sorted, to an unusually distinctive degree, so that the coarser debris forms strips, which are vertical wedges in profile. These develop between adjacent freezing centres that now appear as islands of finer sediment, and each strip becomes one edge of a polygon where the islands of freezing sediment are closely packed to occupy the entire area.

The sorting has been by ice-crystal growth within the soil. There is no doubt that the movements are real, because they have been measured at this particular site by repeated observations of stones identified by paint marks, in a classic piece of diligent fieldwork. However, there is still some mystery over the actual mechanism behind the stones' movement. It can be that ice crystals have grown from random freezing centres during the annual freeze, and have pushed the larger stones towards the zones midway between adjacent centres. Or it can be that the soil fines are squeezed upwards during the thaw, so that the larger stones are displaced laterally and then settle along their polygonal lines. Neither process is easy to conceive, but there is good evidence that both do occur at certain sites. This means that stone polygons are yet another example of a polygenetic landform, whereby different processes have led to the same end. This adds to the confusion, but also to the fascination of geological science.

The Trans-Alaska oil pipeline was one of the great construction feats of the 1970s, and it still continues to carry 250,000 tonnes of 'black gold' every day from the oilfields around Prudhoe Bay to the tanker port at Valdez. Most of the route is across terrain with permafrost, so the pipeline could not simply be buried. Its crude oil has to be kept at temperatures of 40–60°C, as it becomes too viscous if it is any cooler. The challenge for the pipeline is that its own warmth would thaw the permafrost around a buried pipe. It would turn stable frozen ground into soggy and unstable bog, thereby increasing the risk of shallow slope failure or ground subsidence, either of which could rupture the pipeline, with disastrous consequences for the fragile Arctic environment.

In such terrains, conservation of the permafrost is essential to ensure any sort of ground and structure stability. So the pipe rests on trestles, each of which has two pile supports. The piles themselves must be kept frozen into the permafrost; any summer warmth conducted downwards could cause catastrophic ground movement. Most of the piles therefore have internal refrigerators. Inside the tops of these thermopiles, ammonia is cooled to a liquid in the heat fins that are exposed to the Alaskan winter. The liquid falls to the toe of the piles, where it vaporises, and its latent heat is taken from the ground, thereby cooling it. Ammonia gas circulates back up inside the piles, and the cycle repeats. No external power is required, though circulation may reduce during the short summers when air temperatures are too high. Thermopiles distinguish this section of the pipeline, in the pass through the Alaska Range, as it lies across permafrost that is only marginally stable and is covered by taiga vegetation with its scattered trees. Normal piles, without cooling, are adequate on thaw-stable gravels and on rock, and also farther north on the Arctic Plain, where temperatures are so low that the permafrost is stable. The elevated pipeline is a magnificent example of geo-engineering, where ground conditions dictated the entire nature of the construction project.

Gold at the end of the rainbow

Klondike remains one of the most evocative words in any language, even when it is really the name of a minor tributary to the Yukon River in northwestern Canada. Its fame dates from the winter of 1897, when it was the target of the world's greatest-ever gold rush, with thousands of stampeders fighting their way through the unforgiving Arctic terrain. The days of scooping pans full of gold from the alluvium of the Klondike River and its tributary, Bonanza Creek, are now long gone, but were outlasted by a very substantial gold-mining industry. Huge floating dredges followed behind the shovels and sluices of the early miners, until 1966, when they too ceased operations.

Nowadays the gold is won from terrace gravels on the flanks of the two main valleys, and the main extraction method is hydraulicking. Powerful monitor pumps blast jets of water onto exposed faces of soil and rock. These wash away the few metres of organic muck cover, and then eat into the gold-bearing gravels, while also thawing any surviving permafrost. The loose sediment is washed into a basin, from where it is loaded into sluices, and these are highly efficient at separating out the gold particles, which range down to dust size. The monitors' water jets are so successful because they also eat into the bedrock of gold-bearing, altered schist, while leaving untouched the hard, unaltered, unmineralized schist beneath. Most of the Klondike gold is not a classical placer orebody, because the gold was deposited by hydrothermal fluids in both the lower layers of the gravels and also the upper few metres of bedrock. Only some of the gold was reworked by erosion from its original source and re-deposited in the classic alluvial placers that were the target of the original gold rush. This explains why there is no mother-lode within the Klondike catchment; hardly any gold lies in the scatter of hydrothermal veins that lace the bedrock schist. It also explains why a small community of miners continues to extract gold from shallow workings in gravel and schist along the bitterly cold valleys of the Klondike.

Defined as the layer of ground that remains permanently frozen, permafrost is widespread in Arctic terrains. Above it lies the active layer of summer thawing, no more than a few metres thick, which supports a cover of tundra vegetation, except where soil is replaced by permanent ice or bare rock. Below the permafrost, the ground is prevented from freezing by geothermal heat from below, and is known as talik. Most solid rock is little influenced by permafrost. However, a porous soil or an unconsolidated sediment has its pore-water turned to ice, which acts as a strong internal cement and thereby renders the frozen ground relatively solid.

Concealed beneath the active layer, there are few places where permafrost and its ground ice can be seen. One accessible site is at Tuktoyaktuk, an isolated Inuit town on the coast of the Mackenzie Delta in northern Canada. A communal ice-cellar has been excavated within the permafrost, at a depth of about six metres. Short tunnels and rooms are reached from the foot of a vertical shaft that is descended by ladder from what looks like a small hut, except that it has no floor. Within the ice-cellar, the walls provide beautifully clean exposures through the permafrost. In this photograph, the ground-ice appears black, as there are no reflections from it, whereas the buff-coloured material is fine-grained glaciofluvial sand. At about 80%, the proportion of ice is unusually high, and is partly due to the growth of ice lenses in the style of those that can develop into pingoes (of which there are many around Tuktoyaktuk). Disjointed recumbent folding of the ice–sand layering may be partly related to glacial drag when the Cordilleran ice sheet extended this far north. The cellar roof is festooned with large hoar-frost crystals that have formed largely from the breath of the occasional visitors, and the floor is a mix of sand and fallen ice crystals. The high proportion of ground-ice exposed in these walls is convincing support for the need to 'conserve the permafrost' in construction projects. Uncontrolled heat loss from a building can destroy its foundation integrity by thawing the ground, and there would not be much left to support the houses of Tuktoyaktuk if their underlying permafrost did not remain cold.

Icebergs of Ilulissat

Midway along the western coast of Greenland, the whole of Disko Bay offers magnificent panoramas of rock and ice. Just south of the town of Ilulissat (*icebergs* in the Greenlandic language), the rocky peninsula of Sermermiut consists of strong gneiss that is 1740 million years old, so not quite as old as some of its neighbours that form much of southern Greenland (and include the second oldest rocks currently identified anywhere on Earth). The icebergs on the left are emerging slowly from the Kangia Ice Fiord. This is among the Arctic's most prolific sources of icebergs, because it is one of the main outlets from the Inland Ice (as Greenlanders call their central icecap). Geophysical surveys have revealed that the sub-glacial floor of the Greenland Icecap includes a deep valley-system feeding out to the head of the Kangia fiord, which itself is 1000 metres deep. Ice therefore flows towards Kangia, and pours out continually into Disko Bay, at a rate of some 20 million tonnes per day. The smaller icebergs soon melt away, but the largest can be hundreds of metres high and wide and consequently survive for longer as each is swept out into the open sea. They ride the current round Disko Bay, then drift out into the Davis Strait to follow a grand loop northwards, before heading south into the Atlantic. It was quite possibly an iceberg from the Kangia Ice Fiord that caught the *Titanic* during that fateful night in 1912.

In the distance, Disko Island has its own small icecap 900 metres above sea level, on top of a thick sequence of Eocene basaltic lava flows. These volcanic rocks formed as a product of the plate divergence that led to the opening of the Atlantic Ocean and also pulled Greenland a little farther away from Canada's Arctic archipelago. Disko Island is 50 km from Ilulissat, but is clearly visible through the clean Arctic air. Combine this with the minimal plant cover and long summer days, and it's easy to see why Greenland is an absolute delight for geologists.

Well-known as one of the few places where a divergent, constructive, plate boundary can be seen on a land surface, Iceland boasts various locations where it is claimed, albeit fairly loosely, that a person can stand with one foot on the Eurasian Plate and the other on the North American Plate. The Myvatn area, in northeastern Iceland is one of the best divergence sites, and its finest feature is the Grjótagjá fissure out in the old lava fields to the east of the lake. This is a strikingly beautiful open fissure that is constantly increasing in width. Grjótagjá lies along one edge of a small down-faulted block known as a graben. This is itself just one of the multiple dilation features along a zone that is some few kilometres wide and can be identified as the currently active line of the plate boundary. The fissure width is due partly to the plate divergence and partly to rotation of the block of lavas on the right, outwards and downwards into the graben.

The water table at this geothermal hot-spot is about ten metres below surface level, and deep pools within the fissure are fed by hot water that is circulating slowly from depth through dilation fractures in the basalt. When this photograph was taken, the pool temperature was around 60°C, but it has since become cooler by nearly 20°C, so that nowadays the fissure does not normally steam in such dramatic style. In the background, the pyroclastic-rich volcanic cone of Hliðarfjall is no longer active, but the Krafla volcano lies only a little farther to the north along the plate boundary. Though Grjótagjá may grow to greater widths in future years, its position on the plate boundary means that, ultimately, it is doomed to be obliterated. It is quite likely to be filled with new magma derived from below, to become a dyke, or one element within a dyke swarm. Alternatively, it will be buried beneath basalt lavas emanating from some nearby fissure eruption. But until a new eruptive phase generates either of those situations, Grjótagjá will remain a classic geological site that displays divergent plate tectonics in action.

Strokkur

Iceland provided the name for all of the world's geothermal waterspouts. A water spout, or gusher, is known as a *geysir* in Icelandic, and this has evolved into the geological term, geyser. Though it is now almost totally inactive, the real Geysir just about survives in southeastern Iceland's Haukadalur valley, 75 kilometres east of Reykjavik, where it lies at the heart of a hugely popular tourist site. Fortunately for Iceland's tourist industry, the adjacent geyser of Strokkur erupts in grand style at intervals of about five to ten minutes, so every visitor can manage to see plenty of eruptions. Strokkur's activity has been recorded since before 1800, but during the early 1900s it ceased to erupt after its conduit was blocked by ground disturbance during a small earthquake. Subsequently, the blocked conduit in the floor of its pool was cleared out, and since then it has been reliably regular, with eruption fountains usually reaching heights of around 15 metres.

Every geyser erupts when some of its boiling water flashes to steam due to a reduction in hydrostatic pressure caused by the initial steam expansion pushing water out of the top of the conduit when the whole site is heating up. Overflow from the geyser pool signals an impending eruption, which then typically starts with small squirts and fountains that grow progressively taller. However, Strokkur is different. Instead of its steam rising and increasing little by little, a fortuitous characteristic of its conduit morphology causes the steam to rise and emerge as a single mass. This causes the geyser pool to rise into a spectacular turquoise-coloured dome with the bubble of steam visible through its clear water. The dome of water can reach to a metre in height, but it survives only for an instant, before it ruptures with the steam escaping to drive a fountain high into the air. Strokkur is like no other geyser, and is totally mesmerising, with its crowds of visitors waiting for an instant's glimpse of its exploding bubble. This really is geology in action, and less than ten minutes later it all happens again for those who blinked at the wrong moment.

There is always something slightly ethereal about icebergs, which makes this location on the southern coast of Iceland a popular stop for travellers passing along the country's ring road. Breiðamerkürjökull is one of the major outflow glaciers from the huge ice cap of Vatnajökull. The glacier once reached the sea, but now its terminal wall calves into icebergs that float away in the magnificent pro-glacial lake of Jökulsárlón (which means *glacial river lagoon* in Icelandic). This lake is impounded by a redundant terminal moraine, which is breached by a short outlet river into the open sea. Rapid marine erosion of the moraines and the adjacent sandur flats is steadily washing away these unconsolidated sediments that form the narrow strip of land between the sea and the icefield.

The lake has existed only since 1930, when the glacier melted back from the morainic ridge. By 2017 the ongoing ice retreat had left nearly seven kilometres of open water within the lake between glacier and moraine. There is still a steady supply of small icebergs produced by calving of the glacier front. Winds and currents generally sweep these chunks of floating ice down the length of the lake, so that they accumulate close to the outlet river. This passes beneath the ring-road bridge, where a convenient roadside dock allows summer visitors to enjoy an iceberg-spotting boat-trip around the lake.

Jökulsárlón is rather different during the winter (when this was the view from the crest of the moraine barrier). The whole lake freezes over and traps the year's flotilla of icebergs. Their shapes and ice-banding show clearly how many of them had suffered partial melting below the water-line, had therefore become top-heavy, and had rolled over, before being frozen in for the winter. A walk across the thick lake ice is an unforgettable experience, winding a route between towering blocks of glacier ice that are a magical mix of whites and blues. It is unnerving where patches of glass-clear ice look like holes into the lake but can still be walked over, though the enchanting landscape of ice makes the nervous moments worthwhile.

Cutting peat in Mayo

Since time immemorial the blanket bogs of western Ireland have been an invaluable source of peat that provides convenient, welcome and locally available heating in so many houses and farms. The peat can also be regarded as a green resource, in that most of the great bogs are largely a feature of human activities. After the Pleistocene ice sheets finally wasted away, around 15,000 years ago, most of Ireland was soon covered by dense forests of deciduous and coniferous species. It was mainly the Neolithic farmers who cleared large tracts of that early forest, and, once devoid of trees, the soil was rapidly leached by rainfall that washed away much of its contained nutrients. Furthermore, leached minerals of iron and manganese were re-deposited as hard-pan lower in the soil profile, thereby hindering the downward drainage of the land. Heathers and rushes grew on the poor, leached, acidic and saturated soils, and then rotted down into an ever-thickening layer of peat. Trees were choked, and Neolithic walls were buried beneath the encroaching peat; occasionally the remains of both are exposed in the modern peat diggings. Much of Ireland's peat dates from only about a thousand years ago, and about a third of it has since been lost to more than 300 years of extraction for use as fuel.

When peat is cut it is known as turf, and by Irish tradition it is cut only after St Patrick's Day, when spring winds have started to dry the boglands. Each turf cutter works with a specially designed 'slane'; this resembles a narrow spade except that it has a short extra blade at right angles to the main scoop, so that one drive can cut two sides of a peat block and then have it easily picked up. Cut blocks are thrown up onto the bog's surface, where they are subsequently stacked in order to drain naturally and lose some of their huge initial water content. Partially dried, the turf blocks are then gathered from the bogs and restacked against the farms and houses; there they dry completely, to be ready for fuelling the home fires throughout the next winter.

Cliffs of the Aran Islands

Ice-scraped, wind-swept and salt-sprayed, the Aran Islands provide magnificent landscapes of bare limestone strung out across Galway Bay off the west coast of Ireland. They are the last remnants of an escarpment of the Carboniferous Limestone, west of the larger outcrops on the mainland. Those form the Burren, well known for its limestone pavements that are the signature landform of glaciokarst. Though more extensive, the Burren pavements are nowhere quite as bleak and barren as those on the Arans. The limestone dip is around three degrees, and it is to the southwest. So the scarp face on the islands is along their northeast side, where it was rounded and subdued by Quaternary ice sheets that spread southwards from the hills of Connemara. These great masses of ice were hundreds of metres thick and could flow uphill to over-ride features of the bedrock floor, and they pushed up the limestone scarp to create a staircase of low rock terraces. Dramatic cliffs along the southwest side of the largest island,

Inishmore, are the result of relentless wave erosion that has cut far into the dip slope. Exposed to the full force of Atlantic storms, coastal retreat has been massive, and the cliffs are now 50–80 metres high. This spectacular viewpoint stands just to the east of the well-known stone fort at Dún Aonghasa.

Preparation for taking the photograph demanded a modest site investigation. The cantilevered slab of strong limestone weighs more than 100 tonnes, and it will fail one day in the future when fractures develop across it so that the projecting part can drop clean into the sea 70 metres below. A careful check found no deep fissures, nor any freshly broken rock, so it was safe to assume that the added weight of one person was not going to precipitate immediate failure. When failure does occur, it will probably go unseen during a violent storm, when driving rain, pulsing wind, or perhaps a touch of frost, creates natural forces far in excess of those imposed by a passing photographer's model.

Old Man of Hoy

Recognized by many, but seen by not quite so many, Scotland's Old Man of Hoy is a magnificent example of a sea stack. It stands off the west coast of Hoy, the most southwesterly of the Orkney Islands, so that it is easily seen from the Scrabster to Stromness ferry (here seen on its northbound journey). However, the best view is from the cliff top, at the end of a delightful footpath that climbs steadily from the village of Rackwick.

The Old Man is 137 metres tall, and survives as a mere remnant of the extensive vertical cliffs that are retreating in the face of Atlantic storm waves eating at their bases. It stands less than a hundred metres from the main cliff, linked by the remains of a rock rib that is now draped in collapse debris. The stack is formed in strong Hoy Sandstone, a unit within the Old Red Sandstone that was deposited by rivers around 370 million years ago and now forms much of the Orkney Islands. Vertical jointing and almost horizontal bedding in the sandstone ensure the stack's survival, even when it is so tall and narrow.

Critical to the stack's longevity is a basalt lava that poured from a nearby volcano just before the Hoy sands were deposited in their mountain basin. The lava is only about five metres thick beneath the Old Man, but is now fortuitously exposed at sea level, and its strong rock forms a dark platform that takes the brunt of erosive wave action.

Within the sandstone sequence exposed up the walls of the stack, bedding planes and thin shale bands are etched into almost level ledges, and these make splendid nesting sites for birds. Orkney is justly famous for the tens of thousands of sea birds that nest on many of its spectacular sandstone cliffs. The Old Man of Hoy has relatively few avian residents, but these include northern fulmars, which are notorious for their defensive ploy of vomiting on rock climbers who pass by their nest-sites. Rock ledges are abundant on the Old Man of Hoy, but fulmars and climbers have to compete for the convenience of their level benches.

Thornton Force

Known, appreciated, studied and enjoyed by members of numerous school and college field trips, Thornton Force is the highlight of so many students' mapping exercises around the Ingleton waterfalls, in the Yorkshire Dales of northern England. Just north of the Craven Faults that form the edge of the uplifted Askrigg Block, the Force is a classic geological site because it provides a superb exposure of the unconformity at the base of the Carboniferous rock succession. The lip of the waterfall is formed of Carboniferous Limestone, and the plunge pool is cut into basement rocks of Ordovician age. The much-visited midway ledge has a floor of poor-quality slatey rock, an overhang of limestone, and a backwall of basal conglomerate that contains large and small, rounded boulders of basement sandstone sitting on what was once the foreshore of a Carboniferous sea.

Adjacent to the waterfall, Britain's finest exposure of a buried valley (just off to the left of this view) appears as a plug of grass-covered glacial till that fills the old glaciated valley from Kingsdale. Directly above the buried valley, and immediately upstream of the waterfall, a splendid terminal moraine forms a barrier across the mouth of Kingsdale, and this impounded a meltwater lake in front of the glacier as it wasted away at the end of the last Quarternary cold stage. Eventually, the lake spilled over the lowest point along the crest of the moraine dam and found its way back into the entrenched preglacial valley. Its final descent was a cascade down the old valley-side, and there it crossed the limestone/basement boundary. This was the perfect site for a Niagara-type waterfall with strong rock overlying weak rock. Erosion of the lower slatey rocks proceeded apace and headward erosion undercut the upper limestone, enhancing the steepness of the new falls instead of eroding them away. The amount of post-glacial headward retreat is therefore the distance back from the original valley side-slope (preserved as the till/limestone boundary that passes behind the dark yew tree on the left) as far as the modern waterfall. It is about 30 metres, and that took about 10,000 years. Thornton Force offers geology in action, with stories still to be unravelled from both Carboniferous and Quaternary times.

Pippikin Pot

At the western end of the Yorkshire Dales karst in northern England, the limestone plateau extends into the adjacent counties of Lancashire and Cumbria. Within its depths, a rambling network of cave passages is the longest known in Britain, and is now referred to as the Three Counties Cave System. Most of its underground streams form a dendritic network that drains to a single resurgence at a level some 150 metres beneath the lines of sinkholes swallowing water from the adjacent hills. At depth, these active stream caves intersect the abandoned passages of far older cave systems, which are important because they are the only surviving components relating to earlier landscapes. Furthermore they contain stalagmites that can be dated by measuring the isotope ratios of minute traces of radioactive uranium and its daughter products within their calcite. These dated stalagmites yield data points to help create an absolute time-scale that supports understanding of the evolution of the landscapes, where most ancestral surface features have been lost to erosion.

Pippikin Pot is one of the smaller entrances that provide access to the central part of the Three Counties Cave System. The main high-level trunk passage through its part of the cave enlarges into a section with this fine array of stalagmites. None of these has been dated, because removal of even one of them would be regarded as unacceptable damage to the cave. However, less-imposing stalagmites have been removed from obscure corners of nearby passages, and also from other caves within the Yorkshire Dales. The accumulated age-determination data show that most of the calcite was deposited during the interglacial stages of the Quaternary, whereas there was little or no stalagmite growth when ice covered the area. Calcite stalagmites form by precipitation from mineral-saturated percolation water that drips into open cave passages, and they do not grow underwater. Dating of stalagmites in the multiple levels of cave passages therefore reveals a chronology of the declining water tables that indicate successive valley-floor levels within the evolution of the limestone uplands. Calcite cave deposits might be delightful to look at, but they can also provide a wealth of geological data that records landscapes, climates and environments of the past.

Malham Cove

Set into the Yorkshire Dales of northern England, Malham Cove is an iconic feature of the limestone landscape within the National Park. It lies in Britain's finest karst terrain, and is generally described as a dry 70-metre-high waterfall, which in turn is then commonly thought of as a feature of the karst. But it is not a karst landform. Its origins are complex, and not yet understood completely, though it is almost certain that karst processes played only a minor role, subservient to the surface actions of glaciers and rivers. It is a dry waterfall, which briefly came to life twice during 2015, after 190 years of total inactivity. But the rather meagre cascades of tumbling water, then or at any time in recorded history, are largely irrelevant to the evolution of the entire Cove. The same applies to the underground stream that flows through cave passage to emerge underwater in the pool at the foot of the Cove.

The great rock amphitheatre owes its far larger scale to surface erosion by past generations of rivers and glaciers. With the Watlowes valley above the Cove and another valley below, it is clear that rivers have flowed over the Cove in the past. However, it is difficult to see how any flow matching the size of the Watlowes could account for the far greater width of the Cove wall. Even enhanced by glacial meltwater or by *jökulhlaup* floods (glacier bursts) from an ice sheet that sometimes occupied the Tarn basin upstream of the Cove, fluvial erosion alone appears inadequate to account for the Cove. It is clear that Malham was over-run completely by ice sheets during the Quaternary glaciations. Adjacent landforms, including over-deepening of the valley below the Cove, suggest that a powerful ice stream was active within the Yorkshire Dales ice sheet, and that it probably flowed directly over the Cove. Ice-plucking would then have modified and steepened any step in the landscape along the line of a major fault that lies a short way downstream of the Cove. It seems likely that glacial erosion accounted for a large part of the Cove's development. Dry waterfall, cave resurgence and probably a glacial step, Malham Cove is clearly a complex landform, but it does look impressive for the many visitors that it attracts today.

Idol Rock at Brimham Rocks

On the high ground overlooking Nidderdale, in the eastern sector of the Yorkshire Dales, Brimham Rocks are a spectacular cluster of sandstone tors that have become a popular visitor attraction. They are formed in the Lower Brimham Grit, a unit of strong deltaic sandstone within the Carboniferous Millstone Grit Group. Individual tors are up to twenty metres tall, and most have highly indented profiles that pick out the major bedding planes, and the cross bedding, within the almost horizontal rock. Idol Rock is no more than ten metres tall, but is distinguished by its deeply undercut profile. It also stands in splendid isolation, away from the main jumble of rocks, and is therefore missed by many visitors.

Clearly, the shaping of the tors has been due to weathering and erosion of the variable beds within the sandstone, but there are some details of the processes that remain open to debate. The rock now forming each tor is, or was, bounded by sub-vertical fractures that initially focussed the weathering and erosion. These joints are ubiquitous within the local grit beds, but may have been widened at Brimham Rocks by some gentle camber-folding, when the underlying shale was squeezed out. If so, displacement of the sandstone blocks was inevitable, and probably included some sliding towards the adjacent Nidderdale. Alternatively, the joint fissures could have been opened by the effects of glacial drag when the plateau was over-run by Quaternary ice sheets that contained ice streams aligned down both Nidderdale and the Vale of York. Differential weathering of the exposed feldspar-rich sandstone beds has then etched the indented profiles of the tors. It remains debatable just how much wind erosion, by blasting with sand grains or ice crystals soon after glacial retreat, might have contributed to the profiling. Idol Rock is one of the few tors with a deeply undercut base that makes it resemble a zeugen, the 'mushroom-rock' landform that is so characteristic of desert erosion. The real story may seem less exotic, but simple chemical weathering by rainwater and mechanical weathering by frost action are likely to have been the dominant processes behind the shaping of Brimham Rocks.

The Mam Tor landslide

Ever since 1977, when the old main road was closed, the slopes of Mam Tor, above Castleton village in the Derbyshire Peak District, have attracted walkers, geologists and engineers to see a textbook example of a major active landslide. A multiple rotational slide has a slip surface curving 30 metres deep, and the head scars on all the failed rock slices create the steps that break the upper road. In contrast, the lower part of the slide is a debris flow, which gives the lower road its wavy profile between the sheared and displaced slide margins. The landslide has been moving for at least 3200 years, as indicated by the radiocarbon age of a tree root within a soil that was buried by the slide debris. Unfortunately this activity was not recognised when the road was built in the early 1800s; it was merely placed on the easiest slope up out of the Hope Valley, which happened to be across the old landslide. Now the entire hillside, along with the road, moves whenever the complex mass of shale and sandstone is saturated by rainfall. On an average

of one year in four, any winter month with rainfall more than 50% above the winter mean, following a wetter than average year, causes the whole slide-mass to creep by more than half a metre. In each drier year, on average three out of every four, movement is no more than a few centimetres.

This classic relationship between slide movement and groundwater pressures points towards the potential for its engineered stabilization by deep drainage. But the costs would be high, and there could never be a guarantee that part of the slide would not move again in very wet conditions. So the road has been abandoned permanently, is relegated to public footpath status, and is now widely appreciated as a classic geological locality. This photograph dates back to the 1980s, since when the displacements along the upper road have become increasingly conspicuous. Mam Tor simply keeps on moving; and it remains a prime example of where not to build a road.

Peak Cavern

The White Peak is the limestone sector of the Peak District National Park in the English Pennines, and its most impressive karst landscapes are located on and in the hills that surround the village of Castleton. Now a popular weekend destination, this little village expanded from its original site beside the stream that flows out from springs just below the huge entrance of Peak Cavern. The cave gathers drainage from a large part of the limestone hills, and a convoluted route can be followed through large and small high-level passages until it reaches this splendid stream cave, nearly a kilometre from daylight and 200 metres beneath the hilltops.

The rounded profile of the passage confirms its phreatic origins, whereby it was initiated and then grew to its present size when it was filled with water. Its limestone floor and roof were therefore eroded by dissolution at almost equal rates, starting from the bedding plane that is clearly recognizable in the notch along each wall. Water filled the cave at that time because it was ponded behind a section of passage in which it flowed uphill to reach its outlet onto the valley floor at Castleton. That, in turn, implies that the cave lay below the level of the contemporary valley floor, prior to it being deepened by surface rivers and Quaternary glaciers. The deepening allowed the cave drainage to find new outlets at lower levels, which are still active today. During that process, the cave was rejuvenated and partially drained, because most of the passages now lie above the new resurgence level, which determines the local water table. This means that most of the cave is now in the vadose zone, defined as lying above the water table, where its streams flow freely under gravity beneath air surfaces. As part of the rejuvenation process, some new passages were developed along deeper routes that followed lower bedding planes within the limestone, whereas some streams incised canyons into the floors of the older caves. However, it is fortunate that much of the ancestral phreatic cave at Peak Cavern was aligned down the gentle limestone dip, so a long section of the modern, vadose stream flows gently and unimpeded along the floor of this lovely old phreatic tunnel.

Landslide at Ainthorpe

The village of Ainthorpe lies on the northern slopes of the North York Moors, in one of the less-visited parts of northern England; bedrock in the area is Jurassic shale. It achieved its five minutes of fame early in 1999 when a road to the village was broken by a quite modest landslide. As is so often the case with landslides on natural hill-slopes, the causes and processes behind the ground movement were complex and multiple. Not far from the village, Coombs Farm has stood for more than 200 years next to a tiny stream that drains a small part of the moors. Then in the 1980s, artificial changes to the plant cover on the slopes above the farm accelerated rainfall run-off and increased the potential flood peaks. Late in 1993, a single flood event overtopped the stream channel and poured water through the farmhouse; the stream poured in through the front door, down the hallway and out through the back door.

To avoid a repeat, a ditch was excavated across the slope above the farmhouse. This captured part of the flood flow of the stream, carried it away from its channel, and let the water flow out over the hillside away from the farm, before re-joining its channel well beyond the farmhouse. Unfortunately, the ditch outlet was directly above a shallow landslide that had been active more than 100 years previously; since then it had blended into the hillside profile, was obscured by plant cover, and had been totally forgotten. Little more than five years after the new ditch was in place, another winter rainfall event dumped water onto the hillside and re-activated the landslide. The farmhouse was untouched, and had not been flooded, but its approach road was broken where it crossed the landslide's head scar. It was only a small-scale feature, but a step nearly a metre high is highly effective at rendering a road unusable.

Coastal erosion on Holderness

Forming the great sweep from Flamborough Head to Spurn Head, the Holderness coast of East Yorkshire is one of England's classic sites of active coastal erosion. South of the chalk cliffs at Bridlington, bedrock is nowhere to be seen along the coast, and Holderness is fringed by an almost unbroken line of low cliffs formed entirely of Quaternary glacial till. Most of the cliffs are 10–15 metres high, but their heights decline to nothing on the southbound approach to Spurn Head, where the unstable coast transforms into an environment of active deposition and re-erosion. North of that transition, the cliffs of soft, weak till are no match for the waves that pound their bases during every high tide. Cliff failures are frequent, as a succession of bite-sized chunks, each of which is a classic rotational failure within the clay-rich till. Their fallen debris is quickly carried away within the southbound longshore drift that is driven by the North Sea's more powerful waves coming in from the north. So the cliff undercutting starts anew, and the next landslide soon follows.

The coastal retreat averages nearly two metres per year, but at any specific site up to five metres of land can be lost in a single landslide during a winter storm. Between the villages of Skipsea and Ulrome, the coast road in this view is being lost to the sea in a succession of slices. Wholesale coastal defence is not an option at Holderness, where farmland values are low, caravans and chalets are simply rolled back, and the few old farmhouses are abandoned to their fate. Key sites with higher combined values, at Barmston, Hornsea, Mappleton, Withernsea and Easington, are protected by barriers of concrete and armour stone. However, these promote increased erosion due to beach starvation just to the south of each barrier. The master plan is to let the coast evolve into a series of great cusps between these artificial 'hard points', and thereby gain some measure of long-term stability. Currently, this seems to be the only sustainable option, though it might seem hard on the road and the chalets of Ulrome that are subjected to the artificially increased erosion within a deepening cusp.

Dudley limestone mine

Outcrops of the Silurian Wenlock Limestone at Dudley, in the English Midlands, are famous for their fossils, which are so abundant that the trilobite *Calymene blumenbachii* is known locally as the Dudley Bug and previously appeared on the County Borough's coat of arms. Quarries and mines created today's bare rock exposures of Wren's Nest Hill, because this limestone was the vital third component (along with ironstone and coal) that supplied the nearby 18th century iron works and turned the Black Country into Britain's industrial heart. High on the Wren's Nest, a massive, inclined mine gallery, known as a stope followed the 45° dip of the lower limestone bed out to daylight. The method of its mining is known as pillar-and-stall. Sections of the limestone were left in place to form pillars that supported the roof across the 'stalls' where the limestone was mined out, and thereby prevented any subsidence of the overlying ground. Working ceased around 1875, and the surface opening is now known as the Seven Sisters, after its line of huge rock pillars. Unfortunately, the less-pure limestone that the miners left to form the unsupported roof-span is prone to rockfalls; in a natural progression, this breakdown evolves towards a more stable arch, above a growing pile of fallen debris.

This photograph was taken during the 1990s, with a person in silhouette at the far end for scale, when fallen blocks across the mine floor were the product of roof failure that was already extensive. Sadly the scene is now much changed. A major roof fall occurred in October 2001, and the whole mine was considered to be seriously unstable. The lower galleries were therefore filled completely with a coarse rock paste that is effectively a very lean concrete. Then in 2004, the daylight gallery that is seen in this photograph was filled, almost to surface level, with 40,000 tonnes of loose, dry, stone aggregate. The grand plan is that, when funds become available for the necessary remedial works, this can all be removed and the roof can be strengthened with rockbolts; so this dramatic feature of industrial archaeology could yet again be seen by visitors.

Rock blasting at Cheddar Gorge

Cut deeply into the Mendip Hills of southern England, Cheddar Gorge is a popular feature within a spectacular landscape. Each year, huge numbers of visitors pass through or stop within the gorge, and most are on the road that follows the gorge floor. Cliffs, crags and slopes loom over both sides of the road. They are all formed in strong Carboniferous limestone, but are natural features of the landscape. All the exposed rock is therefore subject to continual weathering, erosion and degradation, and on such steep slopes these processes inevitably include small-scale rock-falls and landslides. Potential conflict between visitors and rock-falls has to be minimised by ensuring that the gorge walls are sufficiently stable to eliminate, as far as is possible, the rock-fall geohazard. The cliffs on the southern side of the gorge have the limestone bedding dipping into the face, so that bedding-plane slip is not a threat, and the nearly vertical faces are cleaned of small-scale loose-rock by rope-access geologists on an annual basis.

In contrast, the northern side of the gorge has the limestone bedding dipping out from the face, so that larger landslides are possible where beds of rock can slide over the bedding-plane weaknesses. Over geological time such landslides have created a profile less steep than that of the cliffs on the south side. This shoulder of rock was identified as sliding slowly above a steep bedding plane such that vertical joints behind it were becoming wider. Its limestone was already too broken to be anchored effectively, so it had to be removed before it fell down. The only practical means of doing this was by use of explosives, which was done successfully in two stages of blasting, while the road below was closed to everyone. The photograph is of the first blast. An excessive amount of fly-rock was fully anticipated because the blasted ground was already so fissured that loose blocks of surface rock were simply blown away. A second blast was a less-spectacular, textbook event that produced very little fly-rock, but left a clean, stable profile on newly-exposed sections of the inclined bedding plane and a vertical joint. So this is now another part of the Cheddar Gorge cliffs rendered safe for legions of visitors.

Folds of Millook Haven

Of the many fold structures exposed in the coastal cliffs of Cornwall, those at Millook Haven, near Bude, rank among the more spectacular. The rocks are thinly bedded sandstones and shales of the Upper Carboniferous Culm succession. This sequence of rocks takes its name from culm, the local term for the crushed coal that occurs as thin seams interbedded with the dominant shales and sandstones. These rocks were all formed more than 300 million years ago when the part of the Earth's crust that now underlies Britain was situated close to the Equator. Continents lay to both north and south, with Britain lying within a vast intermediate zone of lowlands, coastal swamps and shallow seas. This entire basin was continually subsiding, so great piles of sediment were steadily accumulating within it. The swamps were the key feature, because they were shrouded in huge rain forests of dense vegetation extending across much of what is now northern Europe. Every so often the rate of tectonic subsidence increased such that the swamps and their forests sank below sea level, and were buried beneath swathes of deltaic sediment carried into the basin by rivers flowing

from the uplands along both northern and southern margins. Those great sheets of buried plant debris were compressed to form coal seams, and the deltaic sediments that buried them became the sandstones and shales of the Coal Measures. Crustal plates evolve with time, and soon after the coal seams were formed they were disturbed by am orogenic phase related to plate convergence. Most of Britain's coal seams lay on the flank of the northern continent, and were gently folded into the structures preserved in the coalfields.

Meanwhile, southern England was caught within a major zone of convergence, where the rocks were compressed and crumpled on a grand scale. Compression turned the coal into high-grade anthracite, but also crushed it to create the culm, and left its thin seams in a series of steep and complex folds. Thin, broken and contorted, the culm is so difficult to mine that it has only ever been extracted on a small scale at a handful of sites. However, the intense folding of the entire Culm succession has provided some remarkable geological structures that are now beautifully exposed in these coastal cliffs.

Preikestolen

One of Norway's best-known landmarks is Preikestolen (which translates as The Pulpit Rock), a granite lookout high above Lysefjord among the mountains inland from the city of Stavanger. Hugely popular with visitors prepared to walk the four rugged kilometres from the nearest road, its nearly flat top, some 25 metres square, offers wonderful views. It has no guard-rails (in line with Norway's admirable policy of not defacing natural features with inappropriate safety measures), so many visitors like to sit right on the edge, which is a clear 604 metres above the cold waters of the fjord. The rock is Proterozoic granite, containing both hornblende and biotite, with a scatter of large white feldspar crystals and small xenoliths of dark rock that were subsumed within the intruding magma. Strong and with few fractures, this is granite at its best, and its cliffs that gleam brightly in the sunshine give the name to Lysefjord, which translates as the Fjord of Light.

Preikestolen was left standing so prominently by the Quaternary glaciers that deepened the fjord and trimmed back its walls. The larger share of the erosion was not by the ice-stream grinding the rock down, but was by the ice plucking away large blocks that were already defined by bedrock fractures; this process is now described as ice quarrying. Fortuitously vertical and widely-spaced joints at the future Preikestolen site allowed blocks to be carried away from three sides, thereby leaving the angular spur. To the visitor this appears to stand directly above the edge of the fjord. But, like nearly all great cliffs, it leans away from the vertical, and there is more than 50 metres of off-set within the lower two thirds of the cliff. This does mean that, even with its top end forming a decent overhang, the great rib of rock that forms Preikostolen is seated well back into the overall cliff structure. Despite the slightly ominous fracture that extends across the back of the rock platform, repeated measurements show there is no sign of movement and no prospect of failure within the present geological environment. This is good for Norway's tourist industry, and also for visiting geologists who want to appreciate an exceptional glaciated landform.

Geirangerfjord

Of all the Norwegian fjords (or fiords), Geiranger ranks among the most spectacular. Its narrow twisting course provides an exciting approach on the ferry from Hellesylt or on the cruise ships that visit every summer. Alternatively, its steep roads provide panoramic vistas for overland travellers, either zigzagging down glacial steps from the east, or taking in the view along to the Seven Sisters Waterfall from the northern road (far right in this view). Like all of Norway's fjords, Geiranger was deepened and trimmed by a succession of Quaternary glaciations, and the last event left tall cliffs that were rendered less stable when supporting ice melted away from in front of them.

Rockfalls and landslides are perfectly natural components of slope degradation along the flanks of these glacially over-deepened valleys, and Norwegian villagers know to avoid placing their houses beneath the steeper, taller or less stable cliffs. But rockfalls have special significance when their debris falls directly into deep fiords, because the resultant tsunamis can destroy fjord-side villages many kilometres from the rockfall sites.

The 1934 event at Tafjord, just north of Geiranger, killed 41 people when a giant wave washed over their village. So there is now considerable interest in the Aknes landslide, which is creeping down a steep slope high above Geirangerfjord (round the corner from this view). The mass of rock on the move is thirty times the size of that which landed in Tafjord, and would create a major tsunami if it all came away in a single event. Geiranger village could be hit by a major wave washing far up the slopes at the head of the fjord. But whether Aknes will continue as a slow creep, will break into smaller slides, or will develop into one large slide is almost impossible to predict. The mountain-side at Aknes is too steep and far too large to be stabilized by artificial means involving rock anchors, concrete buttresses or internal drainage. Instead, the landslide is now monitored intensively, notably for ground vibrations and absolute movements. It is hoped that precursors of any large movements can be recognized, and public warning systems are already in place at Geiranger and Hellesylt, where the effects of geology in action could become rather unwelcome.

A huge and gently graded alluvial fan is one of those landforms that really is best seen from the air. The exit of the gully, ravine or valley from the mountains creates the pivot of the fan, which is built outwards by the stream depositing its sediment as it loses both gradient and energy on entering a lowland basin or a wider valley. The key feature is then that the stream continually swings around to change its course as it slips sideways off its own pile of deposited sediment. With enough changes of course, the end result is an almost perfect fan. The larger fans have profile slopes of only a few degrees, which are hardly spectacular when seen from ground level. Steeper fans can occur, but these are usually smaller or more ragged. Grain-size of the fan material is related to its profile; sand and gravel build the gentler fans, whereas cobbles and boulders are major components at the steeper sites. It is easy to see why such debris is commonly described as a fanglomerate.

This striking example has been built out into the massive glaciated trough of Adventdalen on Spitsbergen, the largest island in the Svalbard archipelago, at the edge of the polar pack ice, far to the north of Norway. Though the popular image of Spitsbergen is of icecaps, glaciers and rocky nunataks, Adventdalen is in the snow-shadow behind the coastal hills, and is near enough to the west coast to be warmed by the tail-end of the Gulf Stream, so that its winter snow-pack melts completely each summer. But in this cold land, devoid of protective soil cover, masses of frost-shattered debris are generated each year, to be swept from the hills by floods of springtime snow-melt. All the rock debris is then deposited as soon as the steep mountain streams lose their power when they sweep out onto the flatter floors of the trunk valleys. Adventdalen's great fan is just one of many on Spitsbergen, and its continued growth is a textbook example of sediment aggradation in its purest form.

Gasterntal

A beautiful, steep-walled Alpine valley, the Gasterntal is tucked away behind Kandersteg in Switzerland's Bernese Oberland. Its towering rock-walls and U-shaped profile make it a textbook example of a glaciated trough, and it still has its own Kander Glacier at its upper end, beyond the distant rock shoulders and out of sight in this view. Between alluvial fans that are largely covered with trees, the flat profile of the valley floor belies the vertical extent of alluvial fill of sands and gravels that mask its deeper profile carved into the underlying bedrock. This is a common situation in glaciated valleys, and there is no indication of the thickness of that fill.

Therein lies the fame of the Gasterntal, for it crosses directly above the alignment of the Lötschberg railway tunnel. In 1908 the southbound heading was being excavated from left to right, roughly beneath the far end of the distant meadow in this view. The heading lay 180 metres beneath the meadows on the valley's alluviated floor. There had been some debate over the potential thickness of the alluvium, but the engineers decided that its base could not be deep enough to reach to the tunnel's level, and decided not to go the expense of a trial shaft in the valley floor. Sadly they had not appreciated the common situation of glacial over-deepening, whereby the valley's rock floor could have a reverse gradient rising to its exposure in a downstream gorge (just off the lower left of this view). The tunnel was being excavated by drill-and-blast, and eventually a round of charges in the pilot heading broke through into saturated gravels. An inrush of water and gravel killed 25 miners, filled more than a kilometre of the tunnel, and left a large sinkhole in the valley-floor alluvium. These gravels had filled much of the valley's depth, and subsequent investigation showed them to be at least 220 metres thick. The Lötschberg tunnel was later re-routed into safer ground and is still in use, but should long be remembered for its major disaster. The recently-built base tunnel lies at a far lower level, entirely within solid rock and more than 300 metres beneath the real rock floor of Gasterntal that lies unseen beneath its alluvial fill.

Landslide at Vaiont

Ranking among the greatest civil engineering disasters of modern times, Vaoint's tragedy in 1963 was unusual in that it involved a brand-new reservoir, but did not include any failure of its impounding dam. The landslide scar still dominates this view from the village of Casso, in the Venetian Alps of northern Italy. Below the bare-rock scar, the huge mass of slide debris reaches to hundreds of metres deep, and is now largely covered by pine forest. The dam still stands (off to the right of this view); it was a brilliantly designed, thin concrete arch 266 metres tall, built within a narrow limestone gorge that provided a perfect dam site. Regrettably, the same cannot be said of the reservoir site. Its one wall was a gigantic slab of strong Jurassic limestone overlying a thin clay bed and dipping at around 30° towards the reservoir. Not recognized at the time, the entire hillside of limestone had previously slipped, during the Pleistocene, and had filled the valley below. Post-glacial erosion had then cut a new valley, thereby removing the toe of the old landslide, and leaving the unstable, slipped limestone slab to form the hillside.

Some kind of failure of the hillside then became almost inevitable when the filling of the reservoir raised pore-water pressures throughout the adjacent ground, thereby reducing frictional resistance to sliding. Creeping movement of the hillside was monitored during construction of the dam, but the fatal mistake was to assume that it would continue as slow creep until stability was achieved over a hypothetical, curved slip surface. Such was not to be, because brittle rupture of some key limestone beds triggered an instant failure. The entire hillside slid into the reservoir. This happened so rapidly that it pushed up a wave more than 100 metres high, which overtopped the dam, hurtled through the gorge below and killed more than 2000 people in the town of Longarone. Amazingly, the dam survived almost unscathed. As did the village of Casso, which was above wave level, although two buildings at the lower edge of the village were destroyed by the air-pressure wave that was generated in front of the water wave. Such was the terrifying scale and speed of the Vaiont landslide.

The sinking of Venice

Already a unique and beautiful city, Venice has achieved extra fame from its frequent flooding – because the entire city is slowly sinking. The total amount of subsidence has been small compared with that at some other sinking cities, but its impact is critical because Venice lies almost at sea-level. Total decline of land level relative to sea level has been less than half a metre during the last hundred years. A major cause of the ground subsidence relates to the thick layers of young, poorly-consolidated clay that underlie the city. These are compressing in response to declining pore-water pressure within them. Starting in the late 1940s, that decline was greatly accelerated by large-scale, industrial abstractions of water from adjacent sand aquifers, but this was halted just before 1970 when its harmful effects were recognized. Natural ground compression and tectonic subsidence do continue, but extremely slowly. The major factor now is the rising sea-level. When most of Venice was built, prior to about 1400, sea levels were falling due to the onset of the Little Ice Age, but they have been rising again ever since about 1650. World-wide, sea-levels are continuing to rise, hence increasing the frequency of flooding episodes, and since 1960 each winter has seen more and more *Acqua Alta* events during which much of the city is flooded at high tide.

Because Venice's ground subsidence is almost completely irreversible, and sea-level rise is going to continue, the potential for frequent flooding remains. The short-term response has been to accept the flooding as part of Venetian life. Raised walkways are largely in place through every winter, but are removed each summer. During an *Acqua Alta*, the boardwalks are needed at times of high tide, and they provide access to all key locations; a published map depicts the dry routes across the city. Away from the boardwalks, Venetian residents and visitors alike have to rely upon wellington boots to stay dry when reaching shops, offices and hotels that continue to operate while their ground floors resemble paddling pools. The flooding of Venice will cease only when moveable barriers are completed across the three entrances to the Venetian lagoon and can be closed to keep out the highest tides; work is in progress but will take some years to complete.

Modro Jezero sinkhole

Perhaps unsurprisingly, Modro Jezero is Blue Lake in the Croatian language. This distinctive and beautiful geological feature lies on the outskirts of Imotski, a small town in the limestone karst of southern Croatia. The lake is normally about 300 metres long and 100 metres across. However, its depth varies from 100 metres to nothing at all when it drains out progressively through its permeable limestone walls, until it has a dry floor at the level of the polje visible in the distance. The lake lies inside a gigantic sinkhole that is more than 700 metres by 400 metres in plan at rim level, and reaches about 290 metres deep when the lake drains away. Such a large sinkhole may be described as a tiankeng; this is a Chinese term for 'sky hole', and has now been adopted internationally as the geomorphological term for the very largest karstic sinkholes. Worldwide, tiankengs are few in number, with the best examples crowded within just small parts of China and Papua New Guinea. There is still some debate over their origins by some combination of rock dissolution and collapse. It is likely that a large doline forms at the surface and is deepened by dissolution; at the same time, a cave chamber traversed by an underground river is formed below and becomes taller by progressive roof breakdown; eventually the intervening rock collapses. There may also be subsequent erosion by streams cascading into the open hole. Variations in tiankeng morphology can be ascribed to different combinations and proportions of the four processes that might have operated at each individual site.

Only a kilometre away from Modro Jezero, Crveno Jezero (Red Lake) is another tiankeng, which is slightly narrower but even deeper than its blue cousin; it looks like the result of a massive collapse that is more recent than that of its neighbour, so its near-vertical walls are less degraded. For any geoscientist making a visit to the Croatian coast, the tiankengs of Imotski are well worth a day-trip inland; they challenge commonly-held perceptions about the events, scales and speeds of the geological processes that fashion our planet's landscapes.

By far the largest of Italy's volcanoes, Mount Etna rises more than 3300 metres above the eastern coast of Sicily, and maintains a state of near-continuous eruptive activity in various styles. Its lower half is a broad shield of basaltic lava, and this is capped by a huge composite cone with horizons and cones of cinders inter-bedded with yet more basaltic lavas. Many of the lava flows have emerged during flank eruptions, and numerous old parasitic vents are marked by splendid cinder cones, each formed when lava effusion diminished during the closing stages of a single eruption event. The summit craters are best known for their almost continuous outpourings of gases; charged with fine ash, these create the thin plume of smoke and dust that is normally seen extending down-wind from the volcano's summit. However the summit activity is hugely variable, and can produce either thick columns of ash that reach kilometres high or lava flows that pour down the flanks. Much more unusual are eruptions that include lava fountains. A series of fountaining events occurred in the summer of 2000. Each eruption lasted for only a few hours, and was followed by a few weeks of inactivity.

One night in June, an eruption started well after dusk, announcing itself with the glow of red lava reflected in a huge steam plume. Lava emerged from the Southeast Crater and flowed for more than a kilometre into the Valle del Bove. A tephra plume towered above it and showered neighbouring villages with lapilli ash. Then a lava fountain rose from the crater, increasing in power until it reached a height of about 400 metres. It was an incredible sight, but it died back to nothing after less than 15 minutes, and the entire eruption ceased about five hours after it started. This photograph is not the best, taken on film and through a telephoto lens, but it has a certain rarity value. The fountain on the left and the lava flow on the right are separated by the plume of dark ash being blown directly over the camera position.

Emerging lava on Etna

All volcanoes provide insights to igneous geology in action, but only a very small vent will allow anyone to get up-close and personal with new igneous rock being formed as magma emerges from the ground. On the Italian island of Sicily, Mount Etna is a massive volcano in an almost constant state of activity. Its summit craters are yawning chasms that are generally either emitting smoke or exploding, and are rarely accessible below their rims. However, Etna is also well-known for its numerous short-lived, parasitic vents that are formed where magma escapes, or has escaped, to daylight at lower altitudes. The basaltic magma is so fluid that gases escape before they can accumulate any high pressure, and much of it therefore exudes from these vents in most peaceful style with no accompanying explosions.

This vent was active for just a few weeks on the southern flank of the volcano. It was only about two metres across, so it was relatively approachable; pure radiation heat prevents a close approach to most large lava flows. The basalt had lost much of its heat in its rise to the surface, so it was already rather viscous; it flowed like toffee, at a rate little faster than that of a fairly athletic tortoise. From an original temperature of more than 1000°C, it was cooling rapidly once out in the open air, towards a temperature of perhaps 500°C, losing its bright red colour only about ten metres from the vent. Down-flow, the lava kept moving beneath its grey crust, but it reached only about 100 metres down the hillside. To extend any further, the lava would have required a higher flow rate to overcome the rapid cooling and inevitable solidification. Only the previous day it had flowed towards the camera to create a classic pahoehoe flow, before cooling a little to become more viscous, then blocking its own track and deflecting further flow to its left. On the edge of the red lava, the thin marginal levee of black aa lava is a precursor to the classic aa flow that developed down-slope and off this view. And now it is just another tiny component of the huge mountain that is Etna.

Strombolian eruption

A volcano's activity is described as Strombolian when it produces a repetitive series of discrete and mildly explosive bursts. Each lasts less than a minute and is followed by an interval of quiet that lasts for 10–20 minutes. Delightful small-scale eruptions of this type occur in vents that are fed by basaltic lava of just the right viscosity; it must not flow continuously (as it does in a Hawaiian eruption), but should require only a modest build-up of gas pressure to fragment the lava that has congealed by cooling within the volcano's throat. The type example is of course Stromboli, in the Italian Lipari Islands just north of Sicily, which has maintained its basic style of activity for more than a thousand years. To those in passing ships, Stromboli is known as the Lighthouse of the Mediterranean.

Up close, Strromboli offers one of the world's great natural sights for those prepared for a dusk walk up to the crater rim or a night spent in a sleeping bag on the crest of the cinder slopes. Each eruption hurls glowing lava about 30 metres into the air; blobs of fully molten lava are mixed with fragments of partly cooled crust. Normally,

Stromboli has two or three active vents, but all are safely inside the summit crater, so they are well away from night-time watchers on the rim. The frequent bursts of fountained red lava are truly spectacular at night, whereas in bright sunlight the eruptions have almost no visible colour. Most lava fragments have cooled before they land, so spatter is rare and almost all of the surrounding slopes consist of loose cinders.

Even Stromboli's activity is not perfectly regular. At intervals of a few years, and without any warning, the volcano produces larger eruptions, with cascades of lava bombs that have injured or killed a few watchers on the rim. So sleeping on the rim is now deemed inappropriate. During 2007, a phase of even larger eruptions produced lava flows down the flank, and prompted partial evacuation of the island's villages. After that event, Stromboli became rather more erratic in its behaviour. But it has now settled back into its proper Strombolian style of regular, small eruptions, and is again accessible by visitors on evening walks.

El Chorro Gorge

Along the Mediterranean coast of southern Spain, the mountains inland from Malaga include a series of long ridges formed of strong Jurassic limestones. Rivers from the interior cut through these ridges in a number of spectacular gorges, of which El Chorro is just one. It carries the Rio Guadalhorce, which was responsible for some destructive flooding of Malaga in the early 1900s. Consequently, two dams were constructed across the river, with the intention of controlling potential flooding and also providing hydro-electric power, with one dam at each end of the El Chorro Gorge. Between 1901 and 1905, a footpath was built through the gorge to provide access between the two dam sites, and was incidentally a boon for the local villagers.

This was no ordinary footpath, because much of it was a dramatic feat of construction, built on steel frames bolted to the gorge walls. The gorge has two sections, each formed in the same beds of strong limestone on opposite sides of an anticline, with a wider valley through the fold core. On the southern side the limestone is vertical, and the gorge is at its narrowest with walls that are also just about vertical. The main gorge cuts straight across the vertical limestone, but has deep wall-recesses cut back along some of the weaker beds. The built walkway has to wrap around these recesses, as in this photograph, where the river flows left to right just beyond, and far below, the two visible sections of walkway. When, many years ago, roads around the gorge were improved, the walkway lost its access value, and fell into disuse, until it was re-discovered by adventurous tourists looking for new forms of excitement. But the fragile structure slowly disintegrated and became too dangerous to keep open. A new walkway was then built, with the old one, just beneath, providing temporary access for the workmen. This was opened in 2015, and has become hugely popular for an exciting few hours' hike through what is undoubtedly a truly awe-inspiring limestone gorge.

In the centre of Windhoek, the capital city of Namibia, an open walkway through the middle of the Post Street shopping centre is adorned with a 'hands-on outdoor sculpture exhibit' that is a welcome surprise to any passing geologist. Thirty-one stainless steel plinths, each around a metre tall, support pieces of 'urban art' that appear at first glance to be chunks of rusty rock, each about the size of a football or slightly larger. But these are no ordinary rocks: they are all meteorites, in their natural state, just as they were found, except that they are now welded onto their steel plinths. They are from the Gibeon meteorite shower that landed in southern Namibia in prehistoric times, and have been found scattered across an area of 300 by 100 km of desert south of Mariental. More than a hundred Gibeon meteorites have been found to date, with the largest weighing in at about 650 kg. Those now on display in Windhoek each weigh about 250 kg. They are thought to be surviving fragments of a single meteorite that probably weighed around 25 tonnes when it broke up on entering the Earth's atmosphere. This was a typical iron meteorite. It consists principally of taenite and kamacite, minerals that are alloys of iron and nickel, and are found only in meteorites. They give the Gibeon meteorites a bulk composition of 8% nickel, 0.5% cobalt, and more than 90% iron.

Meteorites rate fairly highly on geological scales of rarity, but Namibia seems to be rather well blessed. This is particularly so with regard to the iron meteorites, which are composed almost entirely of native metals, as opposed to the rather more common stony meteorites that are formed largely of silicate minerals. Not far north of Windhoek, the iron meteorite of Hoba is the world's largest to have survived its crash-landing intact. It weighs more than 60 tonnes, so it remains where it landed, at a nationally protected site. For seeing meteorites on display, Namibia takes some beating.

The Big Hole

In Kimberley, South Africa, there is only one big hole; and that is The Big Hole, right in the middle of town, with its own viewing platform and visitor centre. It's actually a mine, because it was excavated on a kimberlite diamond pipe. And it pre-dates the town, which grew around it to accommodate the miners. The Kimberley Pipe, as it is now known, was the fourth great diamond pipe to be discovered in the Transvaal, following the initial rush of miners to diggings for alluvial diamonds along the valley of the Vaal River. Farmers nearby were the first to find diamonds just lying on the ground in a few isolated patches away from the river. These patches proved to be the weathered outcrops of the kimberlite rock that occurred as giant pipes and were the real source of the diamonds.

Two pipes lay beneath farmland that was owned by the De Beers brothers. The western pipe was the richest of all, and was the first to be worked on a huge scale, entirely by hand-diggings of the crumbly weathered kimberlite. Individual miners' claims were each 10 metres square, and each miner soon had to make his own ropeway connecting to a colleague on the rim of the ever-deepening pit. Worked downwards at different rates, the ground became a dangerous maze of deep holes and unstable towers. The only way forward was by amalgamation, which was orchestrated by Cecil Rhodes, until the whole site was owned by De Beers Consolidated Mines. They continued excavations, still by hand and still with ropeways up to the rim. Reaching 240 metres deep, it became the world's largest hand-dug excavation, until work stopped in 1914, though machine-powered underground mining continued to greater depths. Since then the upper walls of the hole have flared out in the weak Karoo Shales surrounding the upper part of the pipe, but have remained stable in the strong Precambrian rocks beneath; and water now stands 175 metres below rim level. The Big Hole is a monument to Man's endeavour; it yielded 2700 kg of diamonds from 22M tonnes of rock that was dug out using shovels and hard labour.

Sof Omar Cave

Draining away from the Bale Mountains of southern Ethiopia, the Gestro River follows a deeply incised valley, except at one place where it takes an underground short-cut through the Sof Omar Cave. Though this magnificent river cave is the best part of 20 metres high and wide for its entire route through the meander neck, it's the maze of passages off to the side of the river route that makes Sof Omar so special. The entire cave is formed in horizontal beds of Jurassic limestone, but these lie beneath a cap of Tertiary basalt that forms a wide plateau on both sides of the Gestro valley. Development of the cave pre-dates much of the valley's incision. Most of its passages were formed well below the water table, when the plateau surface and valley floor were both at higher levels. Slowly moving water etched out the joints and fractures within the limestone, especially within various individual beds that were just that little bit more prone to chemical attack. With no scouring, no abrasion and little directional control, dissolution picked out every joint or fault within the limestone to create a huge maze of intersecting passages.

Such passage mazes are common in caves that were developed by slowly moving water rather than by fast, cascading streams, but the maze at Sof Omar is more extensive and better developed than most. Where its passages break out into the valley side near the river cave's exit, sunlight streams in to illuminate a succession of horizontal beds and bedding planes and also the numerous vertical joints. Some of the joints have been opened out to form large cave passages cutting through multiple beds. Others remain only as wall niches and narrow fissures in individual beds. Variations in the rock sculpture seem almost endless. In creating the maze of cave passages at Sof Omar, dissolution of the limestone has revealed so many subtle variations within what at first glance appears to be a single unit of homogeneous limestone. It is fortunate that the River Gestro has cut down deeply enough to drain the caves and reveal these fascinating details of the limestone geology.

Rising above the centre of the Afar Desert in the northeastern corner of Ethiopia, Erta Ale is one of the world's most active volcanoes. A shallow summit caldera contains two vents. Each of these is usually a crater about 30–50 metres deep, and at any given time one may be active with lava while the other is relegated to producing clouds of steam and gas. Within recent years the South Crater has been particularly active, and a lava lake has become an almost permanent feature. Its size and level vary. It can occupy the entire crater floor, more than a hundred metres across, with most of the lake covered by a skin of dark, chilled lava that is broken by fissures overflowing with bright red, molten rock. The slowly moving slabs of chilled lava mimic the style of plate tectonics, with red lava seeping out of some cracks while dark, cooler, lava rolls over and sinks along others. At other times the crater contains only a smaller lava pond, with a turbulent surface of dark skin in smaller pieces between boiling and fountaining red fissures.

Long-lived lava lakes are rare (there are less than half a dozen scattered around the world) because they require a continuous supply of heat to prevent them cooling down, skinning over and ceasing activity. The essential heat supply can only be provided by rising magma, and lack of overflow from a lake means that the new magma is being diverted into dyke systems that are evolving beneath the volcano. Such is the case at Erta Ale, where dyke growth is a significant component of the plate divergence as the Arabian and African plates move apart.

At times the South Crater fills to the brim, and lava overflows to spread across the caldera floor. Erta Ale's lava is extremely fluid, and all conceivable details of classic pahoehoe, with flows, toes, channels, levees and small caves, are created and then preserved beautifully in fragile lava. Keeping fresh in the desert climate, this remains almost completely untrodden, because few visitors ever reach this remote spot. The caldera floor is also distinguished by a liberal scattering of Pele's Hair. These delicate wind-blown fibres of basaltic glass have been formed in large quantities during some of Erta Ale's recent eruption events.

Isolated in the Indian Ocean, and politically a part of Yemen, the magnificent flora and fauna on Socotra have led to the island sometimes being known as the African Galapagos. But there is also some exciting geology within this remote piece of land. A high plateau of Cretaceous and Palaeogene limestones dominates the island and supports some impressive karst landscapes, which contain a number of long and large cave systems. The limestone is wrapped around a granite inlier that forms the island's highest peaks, and much of the coast has only a strip of lowland backed by long lines of high cliffs. Near the eastern tip of the northern coast, the limestone cliffs of Aher rise more than 500 metres above a narrow strip of beach and dunes that can be reached only along a rough jeep track.

Aher is a beautiful location, swept by winds and waves straight off the Indian Ocean. For many millennia those waves have driven vast amounts of shell-sand onto the semi-arid beaches, where the winds have then built them into huge dunes; the dune on the right in this photograph is well over a hundred metres tall. The cliff beyond is distinguished by its two gigantic cave entrances. The lower cave has only a single chamber with a narrow fissure beyond, and the upper entrance has not yet been reached by climbers and cavers. There is also a third cave, with its entrance on a bedding plane almost lost in the shadow half way down the cliff from the floor of the lower big cave. This leads into more than a kilometre of ancient phreatic tunnels, some metres high and wide, which still carry floodwater after rare storm events. The floods have scoured channels through thick banks of calcite flowstone of a type that is common in caves of the Middle East; it was all deposited when the Asian monsoons were a major feature of the region's climate during Quaternary interglacial stages prior to the Last Glaciation. That flowstone is key evidence that Socotra has previously had a more active karst environment, when its climate was far wetter than it is now.

The White Desert of Farafra

Egypt's Western Desert is fringed by a string of oasis basins linked by a single road that loops far out towards the west between Cairo and Luxor. One of those basins is Farafra, nearly 100 km across, and set into the low limestone plateau that borders the vast sand seas of the Sahara. Tucked into the basin's western corner, the green oasis of Farafra has a small town alongside its palmery and its various patches of irrigated farmland. The basin's fringing hills are Palaeogene limestone, but its floor is cut down into snow-white chalk of Cretaceous age. Some sand blows across the basin floor, but it is in short supply, so most of it is stacked into climbing dunes along the downwind, southern margin of the basin.

Farafra's White Desert is named after the eroded remnants of the chalk, which now form a scatter of towers and pinnacles standing proud of the golden sands. These remnant blocks of chalk are all that survives from a karst landscape that matured in the wetter climates of the Neogene, several million years ago. Some of the towers reach to 15 metres tall, and appear to be the remnants of a tropical karst comparable to that which now exists in southern China. They were formed by dissolution of the carbonate rock when there was a plentiful supply of rainfall. However, subsequent climatic change heralded the onset of hot-desert conditions, in which the towers can only degrade. Every day, exposed rock is heated by the sun, and then cools every night. The temperature changes cause minuscule expansion and contraction of the surface layers of the rock, which pick out every tiny weakness or fracture, causing innumerable rock chips and blocks to spall away. Jagged profiles have replaced the more-rounded runnels and flutes that are typical of rock towers in an active karst landscape. Surrounding the towers, numerous smaller pinnacles, each only a few metres tall, were also formed by dissolution of the chalk, but their initial erosion took place beneath a soil cover that has since been replaced by the dry sand. Together, the gleaming white towers and pinnacles rising from a sea of golden sand create a visually spectacular and particularly unusual landscape.

Travertine cascades at Pamukkale

Within the highland catchment of the Buyuk Menderes (that's the River Meander, which gave its name to big river bends), in western Turkey, a gleaming white terrace that can be seen from afar is known as the Cotton Castle, or Pamukkale in Turkish. It is a huge bank of pure travertine with an active zone that reaches more than 2000 metres along the hillside and descends more than 100 metres towards the valley floor. Like all the world's great travertine terraces, it has a geothermal source, with mineral-laden water emerging at 36°C from a cluster of springs along a fault zone between Palaeozoic marbles and Neogene sediments. The terraces have been built by 14,000 years of post-Pleistocene deposition of calcite that has built rimstone dams around numerous pools. However, most of the terraces are no longer active because regional over-abstraction of groundwater, mainly for agricultural use, has greatly reduced the flows from the springs.

For a thousand years the spa town of Hierapolis stood beside the terraces, until it was abandoned in the 14th century. The site was re-born in modern times, to become a major tourist attraction. Bathing in the pools was possible into the 1980s, and visitors were allowed to climb over the dry terraces until some years later. A monumental act of environmental desecration took place during the 1950s, when a road was cut right up the centre of the terraces (it's marked by the oblique line of dark vegetation in the background of this view, which dates from the 1970s). Fortunately, the road was closed permanently in the 1990s, and several low concrete dams were built across it. All of the spring-water outflow is now controlled, and includes some output that is directed down the old road. With total travertine deposition across the hillside reaching several tonnes per day, the entire road and its dams are now coated with layers of pristine white calcite. Bathing is allowed again, but only in these new pools, which many tourists believe to be natural features, because the concrete dams have become totally obscured. Adjacent to them, the natural terraces are refreshed periodically with directed water flows, and they have now lost their pockets of invasive plants (which are visible in this 1970s photograph), and once again they look magnificent. Pamukkale has seen a commendable exercise in managed environmental recovery.

Lying just south of the geographical centre of Turkey, the fretted landscape of Cappadocia forms a justifiably famous tourist destination, within which Zemi Valley is one of the less-visited sites. Cappadocia's fantastical landscapes have all been carved out of pyroclastic debris that was produced about 11.2 million years ago from a nearby series of volcanoes. Eruptions continued after that until a caldera collapse occurred less than a million years ago, and the last-known minor eruption was in 6880 BC. Although some of the volcanic material was airfall ash, most of it was ejected as incandescent clouds that formed pyroclastic flows. The debris was still hot when it settled on the ground, so much of it was welded into stronger material known as ignimbrite. Most of the magma was rhyolitic, which froths up to produce pumice, and the variable extent of its welding accounts for differences seen in the towers and pinnacles of Cappadocia's eroded landscapes.

Locally, the volcanic sequence locally reaches 400 metres in thickness, and subsequent rain-wash, gullying and fluvial erosion have left the numerous rock towers that distinguish the area.

The generally weak ignimbrite is easily excavated with hand tools, and the main valley close to Goreme is famous for its many churches, each of which is carved out within an individual tower; some have beautifully painted interior walls and ceilings. Elsewhere there are entire underground towns, with multiple levels of rooms and galleries carved beneath plateaus of the ignimbrite; their purpose was probably for refuge and defence during violent times around a thousand years ago. Other areas of towers remain in their pristine state, including those in the Zemi Valley, just a kilometre from Goreme, but tucked away down a rough track. The Zemi towers are up to 30 metres tall, with vertical sides beneath a protective caprock. Like those in the more accessible Zelve Valley, the darker caprock is frequently described as basalt lava, but it is actually nothing more than a stronger, better-welded, pyroclastic flow deposit within the ignimbrite sequence. Even though there are many sites that are all so different within this spectacular volcanic terrain, Zemi is a highlight well worth searching out by any geologists visiting Cappadocia.

Sandstone colours at Petra

The Nabataean monuments of Petra, carved into the sandstone cliffs of southern Jordan, are among the world's most spectacular sights. They achieved fame after Johann Burkhardt became the first modern traveller to see them, in 1812. Though described so elegantly as 'the rose-red city half as old as time' by the English ecclesiastical poet John Burgon in 1845, the features visible at Petra today were never a city, and its sandstone is not rose-red. Today's Petra is essentially a necropolis; the spectacular cliff carvings are the facades of tombs that were cut into the rock behind them, including the Silk Tomb shown in this photograph. The rose-red colouring is merely a result of weathering on long-exposed faces.

The Cambrian Umm Ishrin Sandstone is a massive, coarse-grained, fluviatile arkose, in that it is composed of grains of both quartz and feldspar. Its matrix cement includes a large proportion of iron minerals that are brown, red, orange, yellow and black varieties of hematite, limonite and goethite. These create the spectacular colour bands that are essentially large and complex varieties of Liesegang rings. Named after the German colloid chemist who first described their rhythmic precipitation in 1897, it was Wilhelm Ostwald who, in the following year, suggested an origin due to quasi-periodic precipitation. Petra's colour bands were formed during diagenesis, when sand-grade sediment was transformed into sandstone rock, largely by deposition of natural mineral cements in the pore-spaces between the sand grains. Differently coloured iron minerals were deposited in concentric zones, reflecting critical oxidation states within slowly moving groundwater. Patterns relate to multiple phases of pulsed water seepage, with younger oxidation fronts cutting across older sequences of colour bands. All were superimposed on subtle patterns already created by depositional cross bedding in the original sandy sediments. The beautiful colour-banding is seen only on the cut faces inside the rock tombs and on just some of the more-sheltered facades. Elsewhere, weathering has changed the many colours on exposed rock-faces into the uniform rose-red of poetic renown. The Silk Tomb facade was cut back into unweathered sandstone about 2000 years ago; whereas its glorious colours still survive on the facade's lower part, the upper section is already losing its bright colours to the incessant ravages and degradation of rock weathering.

Sap Bani Khamis cliff village

If ever geology can be credited as the dominant influence in town planning, the village of Sap Bani Khamis must rate as a prime example. The entire village stands on ledges of thinly bedded limestone, between vertical cliffs of massive limestone that rise far above and also descend far below. All are within the Cretaceous sequence that forms the anticlinal Jabal Akhdar range in northern Oman. Now abandoned in favour of a more relaxing environment, some of the ruined houses of Sap Bani Khamis remain visible, beneath the overhang just left of centre in this view (align with the margin arrows). Above them vertical limestone wall rises 100 metres, and the cliff below drops sheer or overhanging for nearly 500 metres into the depths of Wadi Ghul. This is commonly known as the Grand Canyon of Oman, where it is cut into the ramparts of Jabal Shams, the country's highest mountain.

The only access to and from the village is along a particularly airy path that follows ledges and ramps within the canyon wall, and has vertical cliffs both above and below. This remarkable path extends for four kilometres from a gully that provides a route down through the upper bed of limestone. Sap Bani Khamis was the ultimate defensive site, established when there was almost perpetual fighting between neighbouring village tribes. It once held about 15 families, the last of which moved out in the 1970s. Survival of the village depended upon the tiny perennial stream that emerges from a bedding plane at the foot of the upper cliff (directly above the houses in this view). The stream flows for less than 100 metres, cascades over small ledges, and then drops into oblivion over the main overhang. To the right of the stream channel, carefully crafted field terraces, now with bare soil, stand directly above open space that extends far beneath the one strong bed of limestone spanning the void. The magnificent geology of Wadi Ghul provided this fantastic cliff village both with the reasons for its existence and also with some dramatic constraints on its expansion.

The interior of Oman is classic desert terrain where water is a scarce resource for most of the time, though there are exceptions. Endless sand dunes of the infamous Empty Quarter spread from the Saudia Arabia across the unmarked border into Oman. Beyond their northern margin the Hajar Mountains are a wilderness of largely bare rock. Lying in between, huge gravel fans with almost imperceptible gradients emerge from mountain canyons and spread out towards the dunes. These fans are laced with wadis, each down-slope from a rocky canyon and eventually dying out in the great sand sea. Conforming with the definition of their name, these wadis are normally dry.

Nizwa is an ancient town built along both sides of a wadi, with its dry gravel channel approaching a hundred metres wide. Nizwa's fort, the critical feature in ancient times, overlooks the wadi, and right beside it the souks are the markets that form the heart of the town. Stone-built covered souks house the traditional hardware and food sales, whereas the twice-weekly goat souk is held outdoors. Both are crowded in the mornings, and most modern Omanis arrive at the souks in saloon cars and pick-up trucks. The wide floor of the adjacent wadi becomes a huge, temporary car park. The system usually works well.

Upstream of Nizwa, the mountain slopes of limestone and ophiolite have no soil cover that could absorb any rainwater, so run-off is almost instantaneous. After gathering in mountain catchments, storm-water rushes down the lowland wadis as flash floods. Unfortunately, one Thursday morning during March 2014, with no sign of rain, everyone parked their cars in the wadi, as usual. But a localized storm on the mountains nearly 20 km away flooded the wadi with hardly any warning for the many car owners who were by then in the souks. The floodwater came from different mountain tributaries, and an initial small flood-pulse warned some drivers to move their cars. Other drivers were too late and they saw their cars drowned by the main flood-wave. A few hours later, the wadi was dry again, but such is the nature of desert flash-floods.

Namakdan salt cave

Salt is far more soluble in water than are either limestone or gypsum, so mountains formed of salt should contain plenty of caves inside them. However, salt caves are few and far between, simply because the world has only a limited number of salt mountains on which karst landforms can develop. Salt is so soluble in both groundwater and rainwater that it can only survive to form large outcrops in desert environments. The Zagros Mountains of southern Iran have the appropriate geology, with folded sedimentary rocks and a thick salt bed at depth, and also have the necessary dry climate. The Farsi word Namakdan means Salt Mountain, and the Zagros range includes many mountains with that name. All are formed on salt domes that have roots formed by salt diapirs squeezed upwards from their Cambrian source horizon that is nearly 4000 metres down. However, the salt mountains exist only when their diapir roots are active and the salt is rising by about a centimetre per year, thereby faster than the land surface is being lowered by dissolution.

Off the southern coast of Iran, in the Persian Gulf, the island of Qeshm has only one salt mountain, and inevitably it is called Namakdan. In keeping with most other salt domes, it is more than five kilometres across and rises several hundred metres above the surrounding plain. With these dimensions, its meagre rainfall has been enough to develop some superb karst landforms along with a number of caves. Brine streams emerge from the active caves, but there are also many old and abandoned cave passages at various levels within the salt mountain. Within these, salt is re-deposited by brine that seeps through their ceilings to form stalactites and stalagmites with shapes and structures that are more complex than is typical in their calcite equivalents. Crystal growth deforms the stalactites so that they twist away from the vertical, and globular forms abound. There are also crystal growths that extend for a few centimetres out from the stalactites, but are far too thin and delicate to be visible in a photograph of this size. Salt karst is rare, salt caves are very rare, and the salt deposits within them are both bizarre and strangely beautiful.

Not lava, but mud; flowing slowly from the vent of a mud volcano. This geological quirk is created where a mixture of mud, gas and warm water rises to the surface within a sedimentary basin. It resembles a volcanic vent with its intermittent eruptive activity, and the flows from different mud volcanoes vary in their viscosity so that individual ratios of mud to water can mimic the morphology of either basaltic or andesitic lavas. An excess of methane gas produces perpetually flaming hillsides, and the vents of some mud volcanoes have been known occasionally to develop short-lived explosive phases, when mud is hurled skywards amid flames that can reach fifty metres into the air.

All these features can be seen near Baku, the capital city of Azerbaijan, a small nation that contains more than half of the world's active mud volcanoes. The Baku sites all emanate from a thick sequence of organic-rich, Tertiary sediments that fill the western part of the Caspian basin. They add an extra element of geological interest to a landscape that is also distinguished by some of the world's oldest oilfields, many of which are still producing. Most of the active vents are on mud volcanoes that are a few metres or tens of metres high and wide, with mud that creeps down their sides like miniature pahoehoe flows of basaltic lava, before drying out and producing splendid desiccation cracks. Typical is this vent on the Perekishqul mud volcano, 40 kilometres west of Baku. It is just one of the dozens of small active vents clustered on the gently rounded summit of Perekishqul. Like most others in Azerbaijan, the entire edifice is some kilometres across, and its continually shifting cluster of mini-vents has built a mud mountain hundreds of metres high. Some of the Azeri mud volcanoes have larger vents that produce huge mudflows reaching lengths of some kilometres, and also occasional gas fires on a gigantic scale. However, these large eruptions are infrequent, and it is the elegant morphology of the smaller, permanently active, vents that have most appeal to anyone interested in geology.

The Amber Room

Created in Berlin during the early 1700s, the amber panels that form the walls of the Amber Room were given to the Tsar of Russia by the King of Prussia in 1716. They were subsequently looted by the invading Nazis in 1941, and were then lost, never to be seen again. Their home in Russia had been a room within the Catherine Palace at Tsarskoye Seloe, some 25 kilometres south of St Petersburg. Fortunately, between 1979 and 2003, a complete replica of the Amber Room was made in its original position within the beautifully restored Catherine Palace. This was a massive undertaking. All the amber was brought from the same source area as the original, along the Baltic coast of Kaliningrad, which remains as a part of Russia forming an enclave tucked between Poland and Lithuania.

Based on old photographs, the new copy of the Amber Room is exact in every detail. The room is about nine by ten metres in plan. One wall is largely occupied by windows, but the other three walls have a total of 55 square metres of amber mosaic in eight panels that reach four metres from floor to cornice. Frames around the panels, and bas-relief ornamentation within them, are beautifully carved in solid amber. Pilasters, doorways and window frames are made of wood gilded with gold leaf. The total effect is stunning. Individual pieces of the amber mosaic are mostly 50–100 mm across and are of random shapes, though they have been cut to fit together with perfection. The carved amber is in larger pieces. Colours vary from yellow to red to brown, but there are almost no inclusions in the Amber Room material, and certainly no recognisable fossil insects, as any mineral with inclusions was better used in jewellery or in museum displays. Around six tonnes of the best-quality output from the large opencast mines of Kaliningrad were used to re-construct the Amber Room, though less than half of that made it into the room after the raw material was sliced or carved. Now a part of the tourist trail around St Petersburg, the Amber Room is a must-see for any visiting geologist.

Columns and cascades of pure ice can decorate a cave even more beautifully than their counterparts in calcite or gypsum can ever achieve. Lomonosovskaya is one of a number of caves cut into the low hills of the gypsum karst of the Pinega Valley, away to the northeast of Arkhangelsk in arctic Russia. The cave's spacious galleries and chambers appear to have been formed largely by major flows of meltwater that were generated underneath a warm-based, Quaternary ice-sheet, probably during the final, wasting stage of the Last Glaciation. A series of parallel, neighbouring cave systems was formed in this way, and only small parts of them were subsequently eroded away by further movements of the ice. Today those same cave passages carry smaller underground streams from open sinkholes, and their flows are augmented by dripping water that percolates through from the forest floor not far above the caves.

The area becomes extremely cold during the famously long Russian winters, but the Pinega karst does lie just outside the arctic permafrost zone. So the ground freezes only down to about a metre deep, and remains unfrozen below that. However, freezing-cold air blows through the main cave passages between their open entrances. Groundwater in fissures within the rock above the caves then continues to seep downwards, but freezes to form ice almost as soon as it enters the cold cave air. The result is a fabulous array of long icicles, frozen cascades, and ceilings frosted with giant ice crystals. At the same time, the cave rivers freeze over, creating underground skating rinks. Nearly all these ice features are formed anew during each winter, and then melt away in the following summer. Every year, the ice is fresh and sparkling, and it makes the caves appear exceptionally beautiful. It is unavoidable that a few muddy footprints are left behind by the tiny number of visitors who get to the caves in winter, but the footprints disappear each summer before a new batch of clean ice is formed during the following winter. These are the ultimate in renewable Earth resources.

Demise of the Aral Sea

The ships' graveyards of the Aral Sea provide some of the world's most poignant photographs of a massive environmental disaster. Now straddling the border between Uzbekistan and Kazakhstan, the Aral Sea was until the 1960s a vast inland lake within the Soviet Union. Its input was provided by two great rivers from the east, while its output was by evaporation within its desert setting, together with rare flood overflows to the Caspian Sea. Misguided Soviet "planning" then diverted most of the flow of the two rivers into huge irrigation schemes, primarily to maintain water-hungry cotton fields. The inevitable result was that the Aral Sea continued to evaporate after losing nearly all its inflow, and its open waters were replaced by desert. At opposite ends of the sea, and 400 km apart, the towns of Aralsk and Moynaq changed from thriving ports to bleak desert towns. Their fleets of trawlers and ferries were stranded in shallows and then left in barren desert as the Aral waters literally disappeared from sight; the two ships' graveyards, one near each dead port, became icons of environmental tragedy.

Since this photograph was taken near Aralsk, in 2001, the impact of the scene has diminished, with many of the ships being cut up and removed for their scrap-metal value; and the trickle of visitors have noticed more greenery across the desert vistas as hardy plants slowly colonize the dry former sea-bed. The one positive change has been construction of the Kok-Aral Dam, which was completed in 2005. This has trapped the reduced flow of the northern river, so that a small North Aral Sea is now stabilized. However, its waters have not returned to the dead port of Aralsk, nor to the remains of the ships' graveyard. Whereas the new sea will never reach either site, its waters do now yield a modest fishery harvest. Meanwhile, the far larger southern part of the Aral has no real prospect of recovery within a vast and empty desert; it is already reduced to a few salt lakes, and eventually they too will disappear.

Hunza

Stretching across northern Pakistan, and beyond, the Karakoram is one of the world's greatest mountain ranges, yet is so widely perceived as overshadowed by its neighbouring Himalaya. It has five of the world's 8000-metre summits (as opposed to nine in the Himalaya). Carving a line through the heart of the Karakoram, the Hunza Valley has its floor at an altitude of just 2000 metres, where it passes the icy ramparts of Rakaposi that rise to 7788 metres. With this beautiful backdrop, an isolated community was formerly ruled by the Mir of Hunza. It joined the outside world in 1979, when the Karakoram Highway, linking Pakistan with western China, brought significant change, and a steady stream of tourists, to the Hunza Valley. The straggling village of Baltit is now known as Karimabad, and overlooks an expanse of farmland that yields a profitable annual harvest of apricots. The fields of Hunza spread across a broad alluvial terrace that is perched more than 50 metres above the main river.

Like nearly all of the great terraces in the Karakoram and Himalayan valleys, this one owes its origins to a major landslide. Within these youthful and still actively growing mountain terrains, landslides are not rare events, but form a significant component of slope degradation. Debris from the larger landslides creates barriers within the valleys, impounding large lakes behind them. Ultimately, the lakes fill with sheets of fluvial sediment, and these are left high and dry as grand terraces when the river cuts a trench through both the landslide dam and the upstream lake sediments. Subsequently, each terrace gains a village or a town, as it offers the best agricultural land within the mountainous terrain. Karimabad's landslide and subsequent lake and terrace development were prehistoric events in the Hunza Valley, but just nine kilometres further upstream a landslide in 2010 replicated the geological story. This new landslide buried the village of Attabad, and impounded a lake 20 km long and 100 metres deep behind the debris dam. Many villagers had to relocate, the Karakoram Highway has been rebuilt, and the lake is now a stable feature of the landscape. Eventually it will fill with sediment, and will create a new terrace for cultivation by a future generation of apricot farmers.

Caves of Ellora

These are not natural caves. They were carved by Man, some 1200 years ago, out of solid basalt, and they are truly amazing in terms of the enormous amount of effort that went into their creation. They lie near Aurangabad in western India, and the basalt they are cut into is part of the massive Deccan Traps, the pile of Cretaceous lavas more than a kilometre thick that poured from vents developed over a mantle-plume hot-spot. Like any basalt, this rock is hard, dense and tough, yet the Hindu masons of years ago chose to work in the strongest of the fine-grained lava, avoiding weaknesses that could lie within the vesicular horizons. Archaeologists claim that it was a good rock to work with, because it hardened on exposure to the atmosphere, and was a little softer when fresh. Nevertheless it was a labour of love (or perhaps enforced slavery) to excavate the dozens of caves in the basalt cliff at Ellora (and also the older caves cut into the cliffs around an incised meander at nearby Ajanta).

Whereas most of the Ellora caves reach some 20–30 metres back into the rock, creating single rooms with walls carved into statues and shrines, the largest Ellora site, the Kailasha Temple, is almost open to the sky. Instead of cutting a cave, the labourers of old cut three trenches that formed a loop back into the hillside. This loop reached 90 metres in, and was 35 metres deep at the back. The block of basalt hillside that remained inside the square-cut loop was then carved into the temple that is seen in the photograph. Some idea of scale is given by the sari-clad lady in the doorway. The whole structure, along with all its intricate ornamentation, was carved from the solid rock, *in situ*; the original ground surface lies in profile across the top of the rock wall beyond the temple. It probably took 7000 labourers more than 100 years to excavate 200,000 tonnes of basalt, all by hand, and carve what was left into an utterly extraordinary temple. This is rock sculpture that really does surpass nature.

Confluence on the Ganges

The two main headwater streams of India's great sacred river, the Ganga (known in the West as the Ganges), converge below the small town of Devaprayag. This lies in the southern foothills of the Garhwal Himalaya, which rise to an altitude of 7138 metres at the summit of Mount Badrinath. However, the two Ganges streams differ markedly in character. The western stream (on the left) is the Bhagirathi, which derives much of its flow from the Gangotri Glacier and the extensive icefields on the northern slopes of the Garhwal peaks. In contrast, the eastern stream is the Alaknanda, which drains the Kedarnath and Badrinath valleys that are fed mainly by rainwater and snow-melt on the southern slopes of the same mountains. This accounts for the difference in colour between the two waters. The one is milky and opaque with its suspended load of finely milled rock-flour from the glacier, whereas the other is a darker, clear green because its gathered rainwater lacks the component of fine-silt sediment.

Such a merging of contrasting waters is seen at many mountain sites, but no other carries the immense religious significance of that at Devaprayag. Many regard this confluence as the spiritual head of the Great Mother River that is the Ganga. So pilgrims travel to the small temple that stands on the rock shoulder directly above the convergence of the two rivers. The more devout pilgrims then immerse themselves in the river on the exact boundary of the clear and milky waters. It is essential that they hold onto stout chains anchored to the rock, to ensure that they are not washed away by the powerful current of icy cold water. Included among the pilgrims are many sadhus, the holy men who spend their lives roaming between the sacred sites of the sub-continent. Though the confluence of the rivers is a visual treat for any passing geomorphologist, the vast majority of its visitors see only something on a higher plane than mere geology. Such is one of the many delight of life in India.

Mount Kailas

Lying far out in western Tibet, Mount Kailas can lay claim to be the most sacred mountain in the world. The higher slopes to the Kailas summit, at an altitude of 6638 metres, are always covered in snow, whereas the barren foothills are clear of snow for most of the year. Consequently, Kailas is normally a shining white pyramid standing above the browns and greys of the Tibetan highlands; it is an inspirational sight, which engenders religious awe. Furthermore, Kailas stands at a convergence of watersheds, so that the great rivers of the Indian sub-continent radiate from it; the Indus, the Sutlej, the Karnali (a major tributary to the Ganges) and the Brahmaputra all originate at Kailas. Therefore it is hardly surprising that this beautiful mountain is held sacred by devotees of the Buddhist, Hindu, Jain and Bon religions. It has never been climbed, and almost certainly never will be. Instead, thousands of pilgrims, mainly Buddhist, go to Kailas every summer and complete a kora – a reverential walk around the entire mountain. This takes two or three days for most pilgrims, and crosses a cold, snowy pass nearly 5700 metres above sea level. In the foreground of this photograph, the Chiu Gompa is one of a handful of small, thriving monasteries that stand at auspicious sites in the harsh and barren landscape surrounding the sacred mountain.

With their minds focussed on a higher plane, few pilgrims appreciate that the whole of Mount Kailas is formed of a 2000-metre-thick sequence of Palaeogene conglomerates. Close to horizontal, these strong rocks are stable in tall cliffs, with variations in bed thickness accounting for the profiles of the mountain and adjacent gorges. They are essentially molasse deposits formed by initial erosion of a chain of andesitic volcanoes that had been erupted as a consequence of active plate convergence. Lying just north of the Himalayas, the Kailas Range now stands on one of the terranes that accreted along the southern edge of the Tibetan Plateau. One of the delights of Mount Kailas is that it represents a convergence of India and Tibet, both in spiritual terms and also in the sense of its geological origins.

Gangapurna Glacier

Tucked away on the remote northern side of the Annapurna Range, the Gangapurna Glacier is one of the geomorphological gems of the Nepal Himalayas. It descends steeply as a series of icefalls from a basin just below the summit of Gangapurna, which at 7455 metres high is a significant peak on the high ridge between the summits of Annapurna I and Annapurna III. In about six kilometres of length, the glacier descends nearly 4000 metres, before it melts and wastes away on the floor of the Marsyangdi Valley. The view from slopes high on the far side of the valley is over the flat-roofed houses in the village of Manang, which looks across to the glacier a kilometre away.

Like nearly every glacier in the world, this one is retreating in the face of global warming. In the Little Ice Age of a few hundred years ago, Gangapurna's ice spread out into a miniature piedmont covering the flat valley floor. In such a deep source-valley it is hardly surprising that the glacier gathered rock debris to create conspicuous lateral moraines on each flank. Its extension out into the larger valley then formed the two magnificent ridges of morainic debris, each more than 50 metres high. At the same time, a rather lower and more ragged terminal moraine was pushed up by the glacier's front at its maximum reach. Now the ice has wasted back, but the moraines survive with a lovely pro-glacial lake enclosed between them. In this photograph the lake surface is still almost entirely frozen from the winter; its far end laps against a snow-covered ramp of outwash gravels that extends from the snout of the glacier, which has now retreated until it is almost out of sight. Fortunately, the lateral moraines are made of a till so cohesive that the inside walls have barely degraded since they lost the support that was provided by the ice until it melted away. Before a rough road was engineered through to Manang in 2015, the Gangapurna Glacier was five days' walk from the nearest highway; but it was worth the effort, because moraines like these bring the textbooks to life.

Mount Everest

Most visitors to Everest see the mountain from the southern side, where it is almost obscured behind the Nuptse ridge, so that little more than the summit pyramid appears over the top. Fewer see Everest from the north, but this is where the great mountain is at its most magnificent. Visible in its entirety, the North Face drops 2500 metres to the head of the Rongbuk Glacier. Most of the rocks forming the uppermost cliffs that are largely snow-free are Ordovician limestones; they overlie Cambrian schists, marbles and gneisses similar to those that form the bulk of the high Himalayas. Scene of adventures by George Mallory, Andrew Irvine, Reinhold Messner and many others, the North Ridge extends left of the summit towards the approach route that climbs from the East Rongbuk Glacier (out of sight farther to the left).

In the foreground, the Rongbuk Monastery thrives in a cold and harsh environment. It is the highest monastery in Tibet and is particularly important because it lies only 25 kilometres from Mount Everest, which has a special place within Tibetan Buddhism, where it is known as Chomolungma, Goddess Mother of the World. Beyond the monastery and between the shadows, multiple ridges of till form extensive moraines left by the Rongbuk Glacier in about 500 years of retreat since the Little Ice Age. They offer shelter for the seasonal climbers' base camps, which are now conveniently reachable by a road that was built by the Chinese to facilitate their first ascent in 1960.

After numerous surveys since the first in 1852, Everest's height is now accepted as 8848 metres. Though measured to an accuracy of a tenth of a metre, quoting a tighter number is pointless because it includes about four metres of snow and ice that can vary with the seasons. Short of a rock fall removing the summit, in the style of that on New Zealand's Mount Cook in 1991, there is no on-going reduction of Mount Everest's height by erosion. However, the mountain's height is increasing by about four millimetres per year, according to current best estimates, due to tectonic uplift within the Himalayan zone of plate convergence.

Glacier firn line in Tibet

A flight out of Lhasa in clear weather offers views of the Nyainqentangla Shan, the first mountain chain north of the Himalayas on a terrain within the multiple accretion zone that forms southern Tibet. Within the chain, one isolated peak reaches to about 6500 metres high, and stands over a glacier that is visible in its entirety. The peak's triangular southern face is nearly all bare rock, because most of its snow slides off in avalanches and then accumulates in the cirque beneath. There the older snow layers are compressed into firn and then into ice, which flows off towards the left into a larger and lower bowl. A heavily crevassed icefall remains active over the intervening glacial step. The glacier flows downhill and then becomes dark and dirty beyond its firn line, which is clearly recognizable in this late-summer view. The firn line marks the transition from clean white snow in the permanent accumulation zone, to dirty ice where debris is left on the surface due to ice melting away from around it during the summer season of maximum ablation. This is however a surface feature, because beneath both the clean white snow and the dirty ice there is clean glacier ice that may be a hundred metres thick.

A glacier's equilibrium line, above which it grows and below which it deteriorates, commonly has its visual expression in the firn line; the exception is where any re-freezing of meltwater accounts for ice accumulation within the dirty zone that lies lower down the glacier. On this glacier, a second icefall lies at the head of a steeper section where the wasting ice descends to much lower levels, so that the surface approaching its toe is covered by a thick blanket of melt-out rock debris. The ice beneath this is exposed only in small parts of the broken terminal wall that overlooks a small proglacial lake, where green, rock-flour-laden meltwater is the end product of those initial snow avalanches. From original accumulation to final ablation, each crystal of ice has moved less than five kilometres, and descended nearly one kilometre, in a journey that took perhaps a hundred years. Such is the life and death of a single small glacier, slow-moving, but still one of the more rapid geological processes.

Across the southern half of Mongolia, the Gobi Desert provides an expansive landscape of semi-arid flatlands that are broken by scattered ranges of crags and mountains. Too dry to support the grasslands of the steppes, which are the signature terrain type of Mongolia, the Gobi supports only sparse dry grass on huge expanses of sand and silt between isolated outcrops of bare rock. Bayanzag is known as the Flaming Cliffs because of the bright red of its Upper Cretaceous sandstones; their colour comes alive at sunset, when they look rather more impressive than they do during an overcast afternoon. Thin beds of hard sandstone terrace the softer profiles of dissected cliffs that reach to a hundred metres above the desert floor.

Most Gobi travellers pass through the regional capital, Dalanzadgad, a dusty town that is now modernising after years of sleepy isolation. One early traveller was Roy Chapman Andrews, leading a palaeontological expedition from the American Museum of Natural History. In 1922, he was passing well to the south of Bayanzag, when a local nomad told him of a 'fossil dragon' that could be seen in the cliffs. He diverted and headed for the cliffs, and found Mongolia's first-recorded dinosaur fossils, of the sheep-sized *Protoceratops*, with a skull that could be thought to resemble that of a mythical dragon. In 1923 Edwards returned to Bayanzag to excavate a wealth of dinosaur fossils, but, more significantly, his expedition made the first discovery of a nest of dinosaur eggs. Accompanied by the bones of what later proved to be a feathered dinosaur, these were the major finds that marked Mongolia as one of the world's great fossil locations.

No longer can dinosaur bones be found lying on the surface at Bayanzag, and there is a complete ban on excavations of the dinosaur-rich horizons. Interest has moved farther west to yet-more-dramatic terrains of dissected red cliffs that continue to yield remarkable fossils. Meanwhile Dalanzagdag looks east towards a gigantic new copper mine that is transforming Mongolia's economy, while the red cliffs of Bayanzag have become a popular destination for a short desert hike on the tourist-trail around the Gobi Desert.

Astride the boundary where the Pacific Ocean plate dives beneath the Eurasian continent, the inhabitants of Russia's Kamchatka Peninsula live with a perpetual threat of seismic activity. Annual plate convergence of around 80 mm generates earthquakes along the eastern coast, as well as fuelling the line of active volcanoes that lie just inland. Most of the peninsula is sparsely populated wilderness, but Petropavlosk is a major city that is spread round the shores of Avacha Bay. This bay provides one of the world's finest natural harbours, but it is set into the seismically active eastern coast. The current prediction is that a nearby earthquake of Magnitude 7 will impact the bay and the city with Intensity up to IX about every 60 years. That constitutes a significant threat to the city's buildings, where thousands of people live in apartment blocks that are typically four to six storeys high.

Most of the more modern blocks have been designed and built with internal shear-walls as part of their structure. These are reinforced elements that are capable of resisting the shear forces induced by lateral vibrations during an earthquake; their effect is similar to that of the diagonal bracing that is developed as an external architectural feature on some modern earthquake-proof office blocks. However, some of Petropavlosk's older apartment blocks lack any shear-walls, so these blocks have been retrofitted with shear-resistant elements. External ribs of reinforced concrete are tied together with tensioned steel cables across the roof and through the foundations. These might not be attractive, but they should be effective in preventing the relatively flimsy walls from being pushed over, which would cause the concrete floor slabs to pancake on top of each other. That style of collapse is the cause of innumerable fatalities during earthquake events in many parts of the world. With the hazard of seismic collapse of their apartment blocks now reduced, the residents of Petropavlosk need only fear earthquake-induced landslides within their hilly city, or perhaps an eruption of either of the potentially explosive volcanoes of Koryaksky and Avacha that loom over the back of the city. In geological terms, Petropavlosk is an exciting place to live.

Kamchatka volcanoes

Along the backbone of Russia's Pacific Coast peninsula of Kamchatka, a magnificent line of volcanoes remains active where the oceanic Pacific Plate is being subducted beneath the continental North American Plate, which reaches out to include the eastern corner of Siberia. Consequently, these andesitic volcanoes are characterized by infrequent but violent eruptions, and a pair of them rises directly behind Petropavlovsk, which is the only city on Kamchatka. The two volcanoes are the beautiful cone of Koryaksky, 3456 metres high, in the distance in this photograph, and Avacha, which is just 2741 metres tall in the foreground. Avacha has had a troubled history since a major lateral collapse that occurred about 30,000 years ago. That was in the style of the well-known eruption of America's Mount St Helens during 1980, but was far larger; it produced landslide debris that is up to a hundred metres thick, on which parts of Petropavlovsk now stand. Since then, the collapse scar has almost been filled by a tall new cone, built up of pyroclastic ashes and relatively viscous lavas from recurring explosive eruptions that were interrupted by minor collapses.

Avacha's new cone had its most recent eruption in 1991, when it produced a lava dome of basaltic andesite. This almost completely filled a deep explosion crater that had been left by the previous eruption in 1945. It also produced a short steep flow of highly viscous lava where it overflowed down the cone's side, but this reached a length of only a few hundred metres. The person in this photograph is standing on the rim of the 1945 crater, and the 1991 lava dome in front of him is still hot enough to produce steam all around its rim, ten years after it was formed. The steam is mainly generated from infiltrating rainwater, but active sulphur vents indicate that there is also a deep-seated heat source. Both Avacha and Koryaksky remain hot and potentially very active. So the key question is that of when the activity in either of them will climax in an explosive eruption and perhaps another lateral collapse, either of which events would have dire consequences for the residents of Petropavlovsk.

Mutnovsky caldera

Any interaction between glacial ice and volcanic heat tends to be spectacular. The great caldera on the active Mutnovsky volcano is no exception, but it is rarely visited because it lies on the remote Kamchatka peninsula on the Pacific coast of Russia. Mutnovsky is a complex of four superimposed volcanic edifices reaching altitudes around 2300 metres, with a caldera some two kilometres in diameter developed in its western flank. A glacier that is sourced within the higher, eastern part of the caldera meets its match in a line of fumaroles and geothermal vents beneath the caldera's western wall.

Most of the caldera wall in Mustnovsky is a dramatic cliff of bare rock, but it is breached by a single outlet ravine. This is partly filled with snow and ice, which provides relatively easy access into the volcano. Once inside the caldera, the visitor is greeted by a panorama of conflicting ice and steam. Squeezed between the caldera wall and the advancing ice, there is an abundance of fumaroles, solfataras, boiling mud pools and noisy vents. These are notable for changing their behaviour or emission style, often within an hour or so, because there is such a delicate balance between the rising geothermal heat and the ground's input of glacial meltwater. More fundamental changes are noticed by repeat summer visitors, when an ice-dammed lake of one year can be replaced by a dry gravel plain laced with fumaroles the following year.

The volcano is still active, but most historical observations have been of short-lived explosive events. These have produced some small pyroclastic columns, occasional kilometre-high fumarolic plumes, a few short phases of Strombolian activity and many, small, shallow earthquakes; there may have been some lava extrusion in 1904, but records are equivocal. The one-day hike into the Mutnovsky caldera rates as one of the world's great experiences of nature, with frequent but irregular changes to the thermal activity adding a frisson of excitement and unpredictability to any visit. Whatever the state of the individual vents, there is something almost mystical about fumarolic steam plumes rising from glacier crevasses.

Lahar damage at Toyako

Usu, in southern Hokkaido, is one of Japan's most active volcanoes. Its main structure is a large basaltic edifice standing on the southern rim of the caldera that contains Lake Toya, but all of its activity in historical times has been dacitic and explosive, both from its central vent and also parasitic flank vents. Late in March 2000, earthquake activity indicated a new eruption on the northwestern flank, immediately behind the hot-springs, lakeside resort town of Toyako, and thousands of people were evacuated without delay. The eruption started on schedule, and a cluster of new vents opened, with phreatic eruptions of steam jets laden with fine pyroclastic ash rising hundreds of metres into the air, each lasting from just a few hours to several days. The largest vent built up a steep cone of airfall ash (properly called tephra) and scattered bombs; small pyroclastic surges reached a few hundred metres towards the town.

After more than a week's volcanic activity, heavy rains destabilized the new cone's flanks, and a mudflow of warm ash (known as a lahar)

swept into the town; its material was almost entirely fine-grained airfall tephra. It followed the line of a pre-existing small valley, and reached as far as the shores of Lake Toya, more than a kilometre from its source. About half way along the valley, it met a road bridge, which was overwhelmed because it had been designed to span only a small river and not a massive lahar. Consequently, the intact, concrete road-deck was lifted from its abutments, and was then carried 160 metres down the valley, floating on the dense slurry of the lahar. Along the way, the bridge deck drifted out of the slurry-filled river channel and hit the Sakuragaoka apartment building. The ground floor of the block was already almost completely buried by the lahar debris, such that the bridge took a chunk out of the corner wall of a first-floor apartment, before coming to rest just beyond (where it remains visible on the debris bank now grassed over). 'My upstairs balcony was hit by a passing road bridge' would have been a classic statement on a subsequent insurance claim form.

Mine collapse at Moonkeuk

In South Korea's hill country some 80 kilometres southeast of Seoul, the Somang Hospice was built during 1984 with rows of patients' rooms along three sides of a square central garden. It was a delightful site until May 2008 when most of the garden disappeared into a hole more than 30 metres deep. Fortunately the buildings remained unscathed, and nobody was hurt, but the site was evacuated without delay. Unbeknown to the hospice's developers, their site lay over an old part of the Moonkeuk gold mine. From an entry on the other side of a hill, miners had extracted ore from a sub-vertical vein within the local granite. In the early 1900s, their highest level extended beneath Somang, where a tall cavity (known as a stope) was opened on the vein; this was extended upwards to about 10 metres below ground level. When all its ore had been worked out, that part of the mine was closed off and forgotten, while workings elsewhere were extended to depths of 380 metres.

The miners thought that 10 metres of bedrock granite, even fractured along the line of the vein, was plenty of cover to leave behind to ensure stability of the ground above. It would have been, except that there was soil more than five metres thick above that part of the mine. For nearly 100 years, that abandoned part of the Moonkeuk mine lay forgotten and unseen. Nearly five metres of granite spanned the 6-metre width of the mined cavity. Nobody saw if blocks of this granite roof fell away at various times over the years. Then May 2008 was marked by a torrential rainstorm. Water seeped downwards through the soil, and then down fractures in the underlying granite. More blocks fell away, and the roof of the cavity migrated upwards. Soon the granite roof-span failed; and inevitably the soil cover was doomed. Nine days after the big rainstorm, the ground collapsed. Subsequent rainstorms heralded further soil collapses and the new sinkhole expanded, but did not quite reach the buildings. With the mine workings extending to such great depths, backfilling is barely realistic, so the site has been abandoned. A man-made sinkhole has replaced a hospice.

Huanglong travertine terraces

Beautiful calcite deposits are well-known in limestone caves, but extend to far greater sizes on the ground surface in karst terrains. Set in the glaciated mountains of northern Sichuan, in western China, at an altitude of 3500 metres, Huanglong is a classic example. A stream that is saturated with lime precipitates calcite where it flows over cascades and loses carbon dioxide to the atmosphere. Significantly, the water also loses yet more carbon dioxide to a host of algae and bryophytes that thrive in the turbulence of the cascades. The resultant deposits are in the form of solid, banded travertine, and the cascades are therefore constructive landforms. Though travertine can be deposited from cold water, the Huanglong water source is like so many of the world's other great travertine sites in that it is a fault-guided geothermal spring of slightly warmed water. This rises beneath a moraine of limestone debris within a magnificent alpine karst terrain. Saturated with lime from its journey through thick Permian limestones, the spring-water deposits travertine in a continuous flight of terraces 3500 metres long, that descends 450 metres along a wooded valley. Travertine dams retain 3400 pools of various sizes, many of them displaying pastel shades of colour that are generated by their temperature-sensitive algae. The effect is stunning, and the long middle section of small, yellow terracettes gives the site its name: Huanglong means Yellow Dragon, and the dragon in Chinese mythology is best regarded as a snake.

Comparisons are inevitably controversial, but Huanglong must rate highly among the world's finest travertine locations. Framed by pine forests, which are buried in snow through the winter, it has a visual warmth that is lacking at some of the more open travertine sites, such as at the Mammoth Terraces in America's Yellowstone National Park. Not many westerners visit Huanglong, and the most sensible way to do so is to join one of the four-day bus tours from Chengdu; these also visit the Jiuzhaigou travertine lakes, and provide a memorable experience amid the crowds of Chinese tourists.

The tower karst of Guilin

Classical Chinese artworks have long depicted almost impossibly-steep mountains rising above rice paddies and placid rivers. To foreigners they were commonly considered to be rather fanciful, but this was not the case, because a traditional artists' community was based at Guilin, where the limestone mountains really are dramatic. This is the area, mainly within Guangxi province, that is dominated by the most extensive, most spectacular and most varied tropical karst terrain anywhere in the world.

Prolonged karstic erosion in a wet tropical environment produces a terrain of increasingly isolated and steep peaks separated by giant dolines and closed depressions. The two main landscape variants are best known by their Chinese names, fengcong and fenglin, which are now part of international karst terminology. Fengcong is a terrain of clustered and overlapping hills separated only by deep dolines; it is sometimes described as egg-box topography. In contrast, fenglin has isolated hills scattered across alluvial plains. Western literature has defined the karst types by the hill shapes, so the gentler slopes of cone karst equate roughly with fengcong, whereas the steeper cliffs of tower karst roughly match most fenglin. In reality the situation is more complex, with considerable variation in peak profiles, even before the influences of geological structure and lithology are taken into account. This site near Caoping, alongside the Li River downstream from Guilin, is an area of well-developed fengcong with a chaotic upland karst terrain that is a clustered mass of steep, conical hills. Cutting through this, the narrow alluvial plain of the Li River, in the foreground of this view across the river, has generated dissolutional undercutting, at plain level. Consequently, the adjacent hills have steepened slopes, and diplay profiles that can well be described as towers. However, these are not isolated towers, each completely surrounded by alluvial plain, which is the classic style of fenglin. That style of landscape is famously well-developed in the Yangshuo area, which is another 40 kilometres down the Li River from this site. The Guilin area has so many variations in its fengcong and fenglin landscapes because tropical karst is not a simple morphology. However, it is invariably spectacular, and it has always been a worthy subject for Chinese artists.

Halong Bay

The numerous limestone islands in Halong Bay form one of the world's most beautiful natural landscapes. Hanging off the coast of northern Vietnam between Hanoi and the Chinese frontier, the bay was almost obscured by political barriers until Vietnam's war was firmly in the past, but it is now a World Heritage Site and an essential port-of-call for most visitors to Vietnam. A day's boat trip winding between the islands – there are actually 742 of them – is a positive delight, and also provides an insight to some outstanding geomorphology. Halong Bay is essentially a drowned tower karst that originally had elements of both isolated fenglin towers and coalesced fengcong cones (to use the correct Chinese terms for these landforms). The smaller islands are just single rock towers with precipitous sides evolved from mature fenglin karst, whereas the larger islands are clusters of cones inherited from fragments of a fengcong landscape. Many islands have short caves within them, including Bo Nau Cave, with its entrance-arch framing this classic view across the hills and islets of Bo Hon Island (one of the sites most frequently visited by the hordes of tourist boats that start out from Bai Chay, the western half of Halong City).

A combination of tectonic subsidence and post-glacial sea-level rise has taken the karst plain between the towers to just below sea level, and the marine invasion has created splendid notches within the tidal range on the new islands. Clearly visible in the view from Bo Nau, these contribute to undercutting the limestone cliffs, and subsequent rockfalls maintain the verticality of the limestone towers. Some notches extend back into caves through the limestone hills, and a few of them extend into drowned dolines that are completely surrounded by forested hills. These inland seas (for they are literally that) are technically known as hongs. Dau Be Island has a chain of hongs that are accessible only by low boats that can pass through the linking caves at low tide. Karstic erosion of the limestone and marine erosion of its coastal hills have combined to form a spectacular and beautiful landscape at Halong Bay.

Mount Popa

There are two Mount Popas in the highlands southwest of Mandalay in central Myanmar, or Burma as it used to be known. The original Mount Popa is a substantial stratovolcano formed of basaltic lavas and pyroclastic deposits, along with landslide debris from various flank failures. A minor eruption in 442 BC is remembered in local legends that are regarded as reliable, but most of the volcano's activity was much earlier, and well back into the Pleistocene. From afar, it now appears as a rather unremarkable mountain with a forested shield more than ten kilometres across capped by a small summit cone that overlooks the scar of an ancient lateral collapse.

Near the outer edge of its western flank, another volcanic vent forms a more conspicuous landform where its resistant core has been exhumed by the erosion of softer rocks from around it. This is the crag of Taung Kalat, which is also widely known as Mount Popa, and overlooks the small town of Popa; it is a popular tourist site, whereas few visitors even notice the true Mount Popa six kilometres away to the east. This smaller Mount Popa is a typical volcanic plug, formed of basaltic andesite. Rising some 200 metres above the surrounding land, its bare-rock walls are broken crags that stand close to vertical around much of its perimeter. A remarkably level summit is nearly 100 metres across, and is occupied by a Buddhist monastery with a fine cluster of golden stupas, a small community of guardian monks and huge numbers of macaque monkeys. The approach is a flight of more than 800 steps, which were once maintained by a solitary hermit monk. They have now been rebuilt in marble for the benefit of the many pilgrims and tourists who have to walk up without wearing shoes. It is debatable whether the smaller Mount Popa of Taung Kalat is a parasitic vent within the larger Mount Popa or is the plug of a totally separate and far older volcano. Either way, the imposing crag is an outstanding geological feature that has become an important part of the local religion and culture.

The Batad rice terraces

The northern half of Luzon, the main island in the Philippines, has a steeply dissected mountainous terrain that is spectacular and beautiful. For the many people living in the numerous towns and villages that are scattered through the uplands, rice is the staple food. It is therefore unfortunate that rice can only be grown in the ponded water of rice paddies when there is a serious shortage of flat land. The indigenous people's response has been the construction of terraces that can support veritable stairways of paddy fields on almost impossibly steep mountain sides. The terraces are all formed in the thick, clay-rich regolith that forms a weathered profile on mainly granitic rocks. They were all hand-built by the Ifugao people, and probably originated as terraces for the growing of taro before conversion to rice cultivation. Most are thought to date back nearly 1000 years.

Still without road access, the village of Batad lies in an isolated and remote valley, within its own magnificent amphitheatre tiered with narrow rice terraces. Stability is ensured because each terrace is faced with a thick wall of rough blocks of stone sealed with clay. Around the nearby town of Banaue, the terraces are even more remarkable because they are faced with extremely steep slopes of clay, though these are internally reinforced by carefully placed blocks of stone. The key geological factor for the terraces is that the local clay is dominated by kaolinite, which is the most stable of the clay minerals, creating the least problems with water absorption and subsequent desiccation cracking. However the terraces survive only due to diligent maintenance by the farmers, who ensure that water is always held in the paddies and any cracks are quickly sealed. Currently, the major problem for the terraces is an onslaught of worms. These never thrived in the rice paddies, but have spread from terraces planted with more profitable crops of vegetables. All too easily the worms' burrows can instigate water flow and piping within the clay, which can lead to terrace collapses unless there is constant vigilance and repair work. This is a massive implementation of small-scale engineering geology, and the Ifuago farmers are clearly masters of their craft.

Rock Islands of Palau

Far out in the Pacific Ocean, some 900 kilometres east of the Philippines, the archipelago of Palau lies on the crests of oceanic plate boundary volcanoes. A shallow lagoon some 50 kilometres long has a fringing reef formed atop ramps that drop steeply towards the ocean floor. Rising from the lagoon are two large islands that consist largely of andesitic rocks, and between them lie hundreds of smaller islands formed of limestone. Though clad entirely in lush green forests, these are known as the Rock Islands, and they are familiar to the many divers who visit Palau to enjoy the unbelievably clear waters between them. The limestones are uplifted Miocene and Pliocene reef and lagoonal facies, which are already well lithified into strong rock. Throughout the Quaternary, and notably during the cold stages with lowered sea levels, these were eroded into a fengcong karst terrain of small conical hills with intervening depressions and dolines. Subsequently, they were partially drowned by the rising sea level, creating the islands of today.

Inside some of the islands, there are hong lakes, which are drowned dolines within the karst terrain. A few of these are linked to the open sea by caves that are traversable only by kayak. But the best known of the hongs is one on Mecharchar Island, which is linked to the ocean only by tidal flows of seawater that pass through networks of narrow fissures within the limestone. Consequently, no large fish or marine predators can reach the lake, and its myriad jellyfish have evolved into stingless forms that thrive on a diet of algae growing within the hongs waters. The islands in this photograph are part of the Ngerukewid group, more easily known as the Seventy Islands. Lying out near the eastern edge of the lagoon, these are isolated from the other Rock Islands, and are now a totally protected bio-reserve, allowing no visits, no landings, no sailing, no diving and no swimming on or around them. So they can only be viewed from the air, and appreciated as yet another of our planet's spectacular limestone landscapes.

The Pinnacles of Mulu

Gunung Api is the highest point along a line of tall limestone mountains within the Gunung Mulu National Park on the island of Borneo; it lies within Sawawak, which is a state within Malaysia. The entire surface of the limestone has been fretted into pinnacles by the dissolutional effects of millions of years of rainfall. This extreme terrain, characteristic of strong, cavernous limestone in the hot, wet, forest environment is known as pinnacle karst, with its tall blades of sharp-edged rock surviving between deep tapering fissures that drain into caves beneath. It can also be described by the Chinese term *shilin*, meaning stone forest.

Known simply as The Pinnacles, high on the summit ridge of Gunung Api, these are among the tallest known anywhere. Blades and spires of limestone rise far above the forest canopy, which itself is generally more than 30 metres above the forest floor, though there is no recognizable floor within this highly fissured terrain. The Pinnacles are only one of many spectacular karst features, including some enormous cave systems, that were first documented by a large and prestigious Royal Geographical Society expedition in 1978. The purpose of that expedition was to document the rain forest, on the sandstone mountain of Mulu and also on the limestone mountains that include Api. This was prior to the immediate area's long-term conservation within a new national park, which was established primarily to protect a sample of the Borneo rain forest from the ravages of timber extraction. The karst and the caves of Mulu now lie at the heart of a small tourism industry, which is important in that it gives the national park a viable economy and provides visitors with a rare opportunity to experience the beautiful rain forest. For the more energetic visitor to Mulu, a visit to The Pinnacles is a highlight, though it requires a hard day clambering up nearly a thousand metres of incredibly steep 'trail'. Karst landscapes are hugely influenced by climate, whereby ice-scoured limestone pavements are the morphological opposite of tropical pinnacle karst. That climb to Mulu's Pinnacles reminds the visitors that the very hot and very wet environment is essential in forming these splendid landforms.

Deer Cave

In the rain-forest wilderness that shrouds the fabled island of Borneo, the Gunung Mulu National Park lies tucked away near the northern tip of the island's Sarawak sector. Not only does this great park provide protection for the myriad fauna and flora of the forest, but it funds itself with revenues from some limited visitor facilities, and also extends across a remarkable range of jagged mountains. These are essentially an escarpment of steeply dipping Palaeogene limestone that is more than 2000 metres thick. Furthermore, this limestone is strong, and fractures are widely spaced across its remarkably thick beds, so it is cavernous on a scale that is matched nowhere else in the world.

The caves of Mulu are enormous, and include the world's largest cave chamber and a huge river system with more than 200 kilometres of interconnected passages. They also include Deer Cave, with its giant passage that is now delightfully accessible, with a long boardwalk through the beautiful rain forest and a concrete path into the cave. The main section of Deer Cave is well over 100 metres wide and is 120 metres tall; the camera position for this photograph is half a kilometre in from the entrance, and the person near the far end of the path is dwarfed by the sweeping rock arch. Everything about Deer Cave is on a gigantic scale. The dark patches on the roof, far above the distant person, are bat roosts; every evening, more than two million bats leave their daytime roosts and emerge from the cave in a spectacular bat flight for a night spent hunting insects over the forest.

Deer Cave was formed by a major river draining off the sandstone slopes of Gunung Mulu, probably well over a million years ago. But the main river now takes another route, and even the smaller modern river in the cave has found its way through a parallel passage at a lower level. The cave of Hang Son Doong, in Vietnam, also has a gigantic main passage, and is far longer; but the cross-section dimensions of Deer Cave still support its claim to be the largest cave passage in the world. It is a hidden gem within the immense forests of Borneo.

Wolfe Creek meteorite crater

One of the two finest and most recognizable impact craters on the planet is that at Wolfe Creek in Western Australia (the other being the Barringer crater in northern Arizona). Nearly a kilometre across, sixty metres deep and with an almost perfect raised rim, the Wolfe Creek site has survived with minimal erosion or degradation since it was created around 300,000 years ago. This is due largely to the fact that it lies in the empty desert of Australia's outback, where geomorphic processes are slowed to negligible rates and the landscapes are dotted with many features of remarkable antiquity. Analysis of the crater morphology implies creation by a meteorite that was probably around 15 metres in diameter and weighed about 25,000 tonnes, arriving with an impact velocity probably in excess of 40,000 kph. As such it would have penetrated well into the ground to create the explosion crater by shock waves generated by the immense heat and instantaneous vaporization of some of the bedrock. Surviving fragments show that it was an iron–nickel meteorite.

Country rock is Proterozoic quartzite, together with a capping of laterite, which is almost ubiquitous across the area. Both the bedrock and the laterite crust were flipped up to form the crater rim that still stands some thirty metres above the surrounding plain. The presence of rather more rim material on the crater's southern side (on the right in this view) suggests a slightly oblique impact. Originally the crater was about 150 metres deep, but it is now more than half-filled with blown sand and lacustrine gypsum, into which the central ring of trees is now rooted. Even though the crater is only a few kilometres from the Canning stock route, which has been used by drovers since 1906, the low hill of its rim was never noticed, and the site was discovered only during an aerial survey in 1947. Wolfe Creek is still an isolated site, and the crater is best seen from the air during an excursion flight from the old gold-mining outpost of Hall's Creek, just 100 kilometres to the north, which is itself a very long drive from anywhere.

Folds in Banded Ironstone

In the great Outback of Western Australia, the Hammersley Ranges are red desert mountains composed largely of Banded Iron Formation. Most of these iron-red rocks are almost flat-lying, and therefore form the wide tablelands of the Pilbara, but the Hammersley Gorge, in the northwestern corner of the Karijini National Park, is famed for its fine exposures through a spectacular zone of folding. The colour-banding is provided by the red of the hematite-stained, silica-rich layers. These lie between the darker bands that can be nearly pure iron ore of hematite and martite; both these minerals have the same composition, which is ferric iron oxide, but the martite was formed by alteration of magnetite and thereby retains a cubic structure. Beautifully delineated by the banding, the folds are classic compressional features, including the anticline and the syncline exposed in this natural rock face. The visual impact of both these fold structures is dominated by the concentric curves of the bedded rock sequences. However, there are details, mainly within the axial zones of both folds, where some of the softer beds have been squeezed and deformed around their more rigid neighbours, so that not all the bedding follows parallel curves.

Tectonic deformation at this site has masked sedimentary features that could reveal more about the origins of the Banded Iron Formation. Cyanobacteria are the blue-green algae that were predecessors to stromatolites. They appear to have oxidized, and therefore precipitated, the iron that was previously held in solution within the early Proterozoic seas of 2400 million years ago. But there is still some debate over the causes of the banding, and how much it could have related to climatic cycles when there was a huge extent of sea ice during one of the planet's earliest cold phases that induced glaciations of the contemporary landmasses. Both east and west of the Hammersley Gorge, this rock unit has less silica and much greater proportions of the iron minerals, hence the colossal open-pit mines that extract the valuable oxide ore at Tom Price, Newman and other sites. Besides having massive economic importance, the Banded Iron Formation of the Pilbara offers geologists both problems to solve and sights to behold.

Sandstone hills of the Bungle Bungles

In the semi-arid Outback that extends across the northern part of Western Australia, the range of rounded hills known as the Bungle Bungles forms a distinctive break in a landscape that, for the most part, is remarkably flat and featureless. Their name has no explicable origin, and they are now equally well-known by their Aboriginal name of Purnululu. Individual hills are mostly around 50 metres high, and they are crammed together to form a chaotic topography with an overall relief of around 200 metres. They are all eroded out of near-horizontal, Devonian sandstones that are well-bedded and with widely variable grain-size.

The strikingly rounded, beehive-shaped towers and cones mimic the hill profiles of some tropical karst landforms in limestone, but these are not formed primarily by dissolution of the rock. The sandstone has been weakened by the leaching of its primary cement, and the hills have then been shaped largely by weathering processes as opposed to fluvial erosion. Within a semi-arid climatic environment, running water has minimal effect, and many landforms evolve from joint-guided, angular blocks towards rounded remnants, much in the style of exfoliation and spheroidal weathering. Ayer's Rock (now known as Uluru) is a very-large-scale version of this long-term weathering into rounded profiles, and the Bungle Bungle hills offer the best of the small-scale examples.

The distinctive colour-banding of the sandstone hills is a feature of variations between beds in their rock sequence. The unweathered rock is pale grey, but its surface oxidizes to a zone that is coloured by red and orange iron oxides formed within it. On top of this weathered zone, cyanobacteria (commonly known as blue-green algae) grow over the surface to create a dark surface crust. However, this remains stable only on the bands of finer-grained sandstone that retain some moisture. In contrast, the near-surface zones of the bands of coarser sandstone dry out quickly in the desert heat, and then lose their crust when it dries and crumbles while its algae die. Consequently, the bands of coarser-grained rock weather back to reveal their rich orange colour. The result is a fine example of natural artistry within geological processes.

The Devils Marbles

One of the geological highlights of the great Australian outback lies beside the Stuart Highway some 400 km north of Alice Springs along the route to Darwin. The Devils Marbles (without any apostrophe, and now formally known by the Aboriginal name of Karlu Karlu, which somehow translates as round boulders) are prime examples of exhumed spheroidal weathering in granite. The host rock is Proterozoic biotite granite that is about 1640 million years old. Its weathered crusts contain some iron oxides, which give the rock the vibrant red colours that are especially striking in evening light. A former sandstone cover across the region was eroded away long ago, and since then the granite has been subjected to deep, sub-soil decay in a classic example of spheroidal weathering.

Spacing of joint-sets throughout the granite varies between about two and eight metres. The joints were filled with groundwater, which reacted with the granite's feldspar minerals to form kaolinite clay, with underground weathering occurring largely in zones just a centimetre or so thick inwards from each joint face of the fractured granite. Rock decay penetrated deeper where it could attack from more sides on edges and corners, and thereby progressively reduced rectilinear, jointed blocks to rounded corestones. These were then exposed on the surface by subsequent erosion and removal of the clay, sand and onion-skin rock shells that had developed along the intervening fractures. Some of these splendid corestones are almost perfect spheres, which are rather more marble-like than this highly photogenic pair, and there are hundreds of boulders spread over an area of a few square kilometres. Some corestones are balanced high on plinths, whereas others are stacked in piles; and a few have been split into pairs of hemispheres along clean fractures that have developed under the stresses of repeated thermal expansion and contraction in the hot desert environment. Spheroidal weathering of granite outcrops is widespread across Australia's interior, but this site must be the most spectacular. Although now protected in principle, the Devils Marbles are unguarded at their remote location, but the massive granite boulders are fairly impervious to potential damage by the few visitors who reach them.

The Welcome Nugget

Scattered around the small towns of Ballarat and Bendigo, the goldfields in the Australian state of Victoria are notable for the number of large nuggets that they have yielded. The ultimate source of the Victoria gold lies in the bedrock of Palaeozoic sandstones and in the veins within them. However, the 19th-century miners soon realized that the best payable ore was in the overlying placer deposits. Furthermore, they found that the best pay-streaks were in the alluvium that lies just above rockhead and were along the lines of old stream courses. To follow the floor of a sediment-filled, buried channel beneath the eastern outskirts of Ballarat, the Red Hill Mine had a horizontal heading 54 metres below ground level. Working in this on June 15, 1858, Richard Jeffrey uncovered the massive Welcome Nugget. It was almost pure solid gold, and it weighed in at 62 kg (the plaster replica in this photograph weighs consideably less). Hauled out of the mine, it was exhibited locally, photographed, and then sent to be displayed in London, before it was sold for £10,500 and melted down to produce sovereigns. At current gold prices, its value would be around £2,000,000.

The only larger nugget ever found was the Welcome Stranger, which was uncovered in 1869 from beneath just two centimetres of soil near Moliagul, west of Bendigo. It contained 72 kg of gold, but this was mixed with a lot of quartz, and it lacked the solidness of the pure gold in the Ballarat nugget. Perhaps that explains why the Welcome Stranger was never photographed before it was crushed and smelted. There is no mining in the Victoria goldfields today, but prospectors continue, using hand-held metal-detectors. Early in 2013 a nugget weighing more than five kilograms was found beneath just half a metre of alluvium near Ballarat.

There has long been a degree of mystery surrounding the origin of large gold nuggets. However, some of them contain twinned crystals of gold, which can only have formed at temperatures higher than 350°C. This confirms that the nuggets originate within veins and are then washed out, to be deposited within the placer deposits. An alternative theory, that nuggets were formed by accretion and annealing within the sedimentary environment, appears to lack the support of any firm evidence.

Tongariro volcano

One of the world's finest one-day walks is the Tongariro Crossing in the North Island of New Zealand. Tramping ranks highly among national sporting activities, and there are often hundreds of trampers crossing Tongariro when the weather is clear on a summer Sunday. It can be a colder but more peaceful experience in the months adjacent to winter. From one roadhead to another, the route traverses a saddle between summits within the large and complex Tongariro volcano, and provides spectacular views of a host of cones, craters and exposed volcanic features. Best known are the twin Emerald Lakes, each in its own small crater on the eastern flank, but the skyline is dominated to the south by the larger cone of Ngauruhoe, which carries a snow cap through most of the winter. Its steep, symmetrical, andesitic cone is the youngest component of the Tongariro complex, having been formed entirely within the last 2500 years. It is now one of the more active volcanoes in New Zealand, with a tendency to generate significant pyroclastic eruptions and small lahars at roughly ten-year intervals. In the foreground of this photograph the Tongariro Crossing trail lies along a snow-crested ridge of volcanic ash, which separates two large explosion craters. On the left, the dark far wall of the Red Crater exposes a sub-vertical dyke that originally extended outwards and upwards from the active vent. The dyke is distinguished by the fact that its basaltic magma drained away after a metre-thick layer had been chilled against both of its walls, thereby preserving its structure around an interior void.

Tongariro and its neighbour to the south, Ruapehu, are the largest volcanoes along the convergent plate boundary that underlies New Zealand's North Island. This active line of subduction is responsible for the extensive Taupo Volcanic Zone, wherein it also powers the geothermal sites in the calderas of Rotorua and Lake Taupo. The entire region presents a classic illustration of volcanic geology, and for many visitors the Tongariro Crossing is a highlight in terms of both altitude and geological spectacle.

Pohutu Geyser

There are geothermal areas scattered all around the world, but only four of them have significant geysers among their displays of boiling pools and bubbling mud. America's Yellowstone has the greatest number of geysers, whereas Iceland has only a few, which include the original Geysir after which they are all named. Russia's Kamchatka and New Zealand's Rotorua vie for second place in the geothermal rankings. Currently the finest of Rotorua's geysers is Pohutu, which erupts to heights of about 20 metres, and occasionally much higher. Like all geysers, the timing of its eruption cycle can vary over time and also with seasonal changes in groundwater conditions, but Pohutu is distinctive in that it can spend an unusually large proportion of its cycle hurling water high into the air. Eruptions normally occur at roughly hourly intervals, each with a spectacular full-height fountain lasting for about 20 minutes. Furthermore they are usually accompanied by eruptions of the adjacent Prince of Wales Feathers Geyser, which shoots out a smaller, inclined jet of water at an angle of 30° from the vertical. Perched on massive banks of white siliceous geyserite, this pair of geysers creates a magnificent display, and their location in Whakarewarewa is an added attraction.

Whaka (as it is commonly known) is the thermal village. This has a community of Maori families living in houses built amid the geothermal field, with hot pools, boiling mud and many smaller geysers in their gardens and right next to their homes. Established long before risk assessments and safety regulations were conceived, the site gave the villagers the multiple benefits of district heating, hot water for washing and even steam for cooking. There is no other geothermal field quite like Rotorua. The area is also distinguished by the scale at which its geothermal resources are utilised, notably by large power stations driven by steam from extensive well-fields south of Rotorua. There are also numerous boreholes extracting hot water. Inevitably, these have all combined to over-abstract geothermal heat to the extent that many of the region's geysers have diminished in power or ceased activity entirely. Fortunately, Pohutu continues to erupt, and the thermal village still has its own special heating system.

Moeraki Boulders

On the eastern coast of New Zealand's South Island, a sweep of beach not far north of the village of Moeraki attracts a succession of visitors who come to see its famous collection of boulders when they are exposed at low tide. These are the Moeraki Boulders. Currently, there are about fifty of them, each up to about two metres in diameter and remarkably close to perfectly spherical. They are nodules that have been washed out of the poorly consolidated mudstone of the Palaeogene Moeraki Formation, which forms low cliffs along the head of the beach. Like most nodules, these were formed during diagenesis, when the original sediment was buried and transformed into a sedimentary rock. As part of that process, calcite was deposited in the inter-granular pore-spaces, thereby cementing the fine sediment and turning it into sedimentary rock. Deposition was not uniform, but was concentrated around scattered organic nuclei, where the calcite accumulated preferentially to produce spherical nodules within the otherwise homogeneous mudstone. Moeraki's nodules are also septarian, in that their interiors are laced with shrinkage fractures that were subsequently filled with crystalline calcite. Many of the boulders show traces of the fractures on their exteriors, but fortunately the spherical nodules have strong, densely cemented crusts that keep them intact when first exposed. Ultimately, erosion breaches the crusts, and the interiors then fall apart to reveal yellow calcite within the septarian fractures.

Continuing marine erosion of the cliffs yields a slow but steady supply of new nodules, to replace those that are destroyed by the same wave action farther out on the beach. The intact spherical boulders are the beach's main attraction, but the septarian interiors of broken nodules are also there to be seen, though there are no recognizable fossils preserved inside the boulders. Large Cretaceous nodules that are similarly exposed in cliffs a few kilometres to the south, have yielded fossils, including an almost complete plesiosaur. However, those older versions have the more typical nodule shape, which tends towards ellipsoidal rather than spherical because they grew faster and farther along the bedding direction. Fossiliferous those may be, but they cannot match the sheer spectacle of the Moeraki Boulders, where natural processes have reached close to mathematical perfection.